T0136224

TUNNELS AND UNDERGROUND CITIES: ENGINEERING AND INNOVATION MEET ARCHAEOLOGY, ARCHITECTURE AND ART

PROCEEDINGS OF THE WTC2019 ITA-AITES WORLD TUNNEL CONGRESS, NAPLES, ITALY, 3-9 MAY, 2019

Tunnels and Underground Cities: Engineering and Innovation meet Archaeology, Architecture and Art

Volume 10: Strategic use of underground space for resilient cities

Editors

Daniele Peila
Politecnico di Torino, Italy

Giulia Viggiani
University of Cambridge, UK
Università di Roma "Tor Vergata", Italy

Tarcisio Celestino
University of Sao Paulo, Brasil

CRC Press
Taylor & Francis Group
Boca Raton   London   New York

CRC Press is an imprint of the
Taylor & Francis Group, an **informa** business

A BALKEMA BOOK

Cover illustration:

View of Naples gulf

CRC Press/Balkema is an imprint of the Taylor & Francis Group, an informa business

© 2020 Taylor & Francis Group, London, UK

Typeset by Integra Software Services Pvt. Ltd., Pondicherry, India

Published by: CRC Press/Balkema
        Schipholweg 107C, 2316XC Leiden, The Netherlands
        e-mail: Pub.NL@taylorandfrancis.com
        www.crcpress.com – www.taylorandfrancis.com

ISBN: 978-0-367-46878-1 (Hbk)
ISBN: 978-1-003-03170-3 (eBook)

Tunnels and Underground Cities: Engineering and Innovation meet Archaeology,
Architecture and Art, Volume 10: Strategic use of underground
space for resilient cities – Peila, Viggiani & Celestino (Eds)
© 2020 Taylor & Francis Group, London, ISBN 978-0-367-46878-1

Table of contents

*Tunnels and Underground Cities: Engineering and Innovation meet Archaeology,
Architecture and Art, Volume 10: Strategic use of underground
space for resilient cities – Peila, Viggiani & Celestino (Eds)*
© 2020 Taylor & Francis Group, London, ISBN 978-0-367-46878-1

Preface

The World Tunnel Congress 2019 and the 45th General Assembly of the International Tunnelling and Underground Space Association (ITA), will be held in Naples, Italy next May.

The Italian Tunnelling Society is honored and proud to host this outstanding event of the international tunnelling community.

Hopefully hundreds of experts, engineers, architects, geologists, consultants, contractors, designers, clients, suppliers, manufacturers will come and meet together in Naples to share knowledge, experience and business, enjoying the atmosphere of culture, technology and good living of this historic city, full of marvelous natural, artistic and historical treasures together with new innovative and high standard underground infrastructures.

The city of Naples was the inspirational venue of this conference, starting from the title Tunnels and Underground cities: engineering and innovation meet Archaeology, Architecture and Art.

Naples is a cradle of underground works with an extended network of Greek and Roman tunnels and underground cavities dated to the fourth century BC, but also a vibrant and innovative city boasting a modern and efficient underground transit system, whose stations represent one of the most interesting Italian experiments on the permanent insertion of contemporary artwork in the urban context.

All this has inspired and deeply enriched the scientific contributions received from authors coming from over 50 different countries.

We have entrusted the WTC2019 proceedings to an editorial board of 3 professors skilled in the field of tunneling, engineering, geotechnics and geomechanics of soil and rocks, well known at international level. They have relied on a Scientific Committee made up of 11 Topic Coordinators and more than 100 national and international experts: they have reviewed more than 1.000 abstracts and 750 papers, to end up with the publication of about 670 papers, inserted in this WTC2019 proceedings.

According to the Scientific Board statement we believe these proceedings can be a valuable text in the development of the art and science of engineering and construction of underground works even with reference to the subject matters "Archaeology, Architecture and Art" proposed by the innovative title of the congress, which have "contaminated" and enriched many proceedings' papers.

Andrea Pigorini
SIG President

Renato Casale
Chairman of the Organizing Committee WTC2019

Acknowledgements

REVIEWERS

The Editors wish to express their gratitude to the eleven Topic Coordinators: Lorenzo Brino, Giovanna Cassani, Alessandra De Cesaris, Pietro Jarre, Donato Ludovici, Vittorio Manassero, Matthias Neuenschwander, Moreno Pescara, Enrico Maria Pizzarotti, Tatiana Rotonda, Alessandra Sciotti and all the Scientific Committee members for their effort and valuable time.

SPONSORS

The WTC2019 Organizing Committee and the Editors wish to express their gratitude to the congress sponsors for their help and support.

*Tunnels and Underground Cities: Engineering and Innovation meet Archaeology,
Architecture and Art, Volume 10: Strategic use of underground
space for resilient cities – Peila, Viggiani & Celestino (Eds)
© 2020 Taylor & Francis Group, London, ISBN 978-0-367-46878-1*

WTC 2019 Congress Organization

HONORARY ADVISORY PANEL

Pietro Lunardi, President WTC2001 Milan
Sebastiano Pelizza, ITA Past President 1996-1998
Bruno Pigorini, President WTC1986 Florence

INTERNATIONAL STEERING COMMITTEE

Giuseppe Lunardi, Italy (Coordinator)
Tarcisio Celestino, Brazil (ITA President)
Soren Eskesen, Denmark (ITA Past President)
Alexandre Gomes, Chile (ITA Vice President)
Ruth Haug, Norway (ITA Vice President)
Eric Leca, France (ITA Vice President)
Jenny Yan, China (ITA Vice President)
Felix Amberg, Switzerland
Lars Barbendererder, Germany
Arnold Dix, Australia
Randall Essex, USA
Pekka Nieminen, Finland
Dr Ooi Teik Aun, Malaysia
Chung-Sik Yoo, Korea
Davorin Kolic, Croatia
Olivier Vion, France
Miguel Fernandez-Bollo, Spain (AETOS)
Yann Leblais, France (AFTES)
Johan Mignon, Belgium (ABTUS)
Xavier Roulet, Switzerland (STS)
Joao Bilé Serra, Portugal (CPT)
Martin Bosshard, Switzerland
Luzi R. Gruber, Switzerland

EXECUTIVE COMMITTEE

Renato Casale (Organizing Committee President)
Andrea Pigorini, (SIG President)
Olivier Vion (ITA Executive Director)
Francesco Bellone
Anna Bortolussi
Massimiliano Bringiotti
Ignazio Carbone
Antonello De Risi
Anna Forciniti
Giuseppe M. Gaspari

Giuseppe Lunardi
Daniele Martinelli
Giuseppe Molisso
Daniele Peila
Enrico Maria Pizzarotti
Marco Ranieri

ORGANIZING COMMITTEE

Enrico Luigi Arini
Joseph Attias
Margherita Bellone
Claude Berenguier
Filippo Bonasso
Massimo Concilia
Matteo d'Aloja
Enrico Dal Negro
Gianluca Dati
Giovanni Giacomin
Aniello A. Giamundo
Mario Giovanni Lampiano
Pompeo Levanto
Mario Lodigiani
Maurizio Marchionni
Davide Mardegan
Paolo Mazzalai
Gian Luca Menchini
Alessandro Micheli
Cesare Salvadori
Stelvio Santarelli
Andrea Sciotti
Alberto Selleri
Patrizio Torta
Daniele Vanni

SCIENTIFIC COMMITTEE

Daniele Peila, Italy (Chair)
Giulia Viggiani, Italy (Chair)
Tarcisio Celestino, Brazil (Chair)
Lorenzo Brino, Italy
Giovanna Cassani, Italy
Alessandra De Cesaris, Italy
Pietro Jarre, Italy
Donato Ludovici, Italy
Vittorio Manassero, Italy
Matthias Neuenschwander, Switzerland
Moreno Pescara, Italy
Enrico Maria Pizzarotti, Italy
Tatiana Rotonda, Italy
Alessandra Sciotti, Italy
Han Admiraal, The Netherlands
Luisa Alfieri, Italy

Georgios Anagnostou, Switzerland
Andre Assis, Brazil
Stefano Aversa, Italy
Jonathan Baber, USA
Monica Barbero, Italy
Carlo Bardani, Italy
Mikhail Belenkiy, Russia
Paolo Berry, Italy
Adam Bezuijen, Belgium
Nhu Bilgin, Turkey
Emilio Bilotta, Italy
Nikolai Bobylev, United Kingdom
Romano Borchiellini, Italy
Martin Bosshard, Switzerland
Francesca Bozzano, Italy
Wout Broere, The Netherlands

Domenico Calcaterra, Italy
Carlo Callari, Italy
Luigi Callisto, Italy
Elena Chiriotti, France
Massimo Coli, Italy
Franco Cucchi, Italy
Paolo Cucino, Italy
Stefano De Caro, Italy
Bart De Pauw, Belgium
Michel Deffayet, France
Nicola Della Valle, Spain
Riccardo Dell'Osso, Italy
Claudio Di Prisco, Italy
Arnold Dix, Australia
Amanda Elioff, USA
Carolina Ercolani, Italy
Adriano Fava, Italy
Sebastiano Foti, Italy
Piergiuseppe Froldi, Italy
Brian Fulcher, USA
Stefano Fuoco, Italy
Robert Galler, Austria
Piergiorgio Grasso, Italy
Alessandro Graziani, Italy
Lamberto Griffini, Italy
Eivind Grov, Norway
Zhu Hehua, China
Georgios Kalamaras, Italy
Jurij Karlovsek, Australia
Donald Lamont, United Kingdom
Albino Lembo Fazio, Italy
Roland Leucker, Germany
Stefano Lo Russo, Italy
Sindre Log, USA
Robert Mair, United Kingdom
Alessandro Mandolini, Italy
Francesco Marchese, Italy
Paul Marinos, Greece
Daniele Martinelli, Italy
Antonello Martino, Italy

Alberto Meda, Italy
Davide Merlini, Switzerland
Alessandro Micheli, Italy
Salvatore Miliziano, Italy
Mike Mooney, USA
Alberto Morino, Italy
Martin Muncke, Austria
Nasri Munfah, USA
Bjørn Nilsen, Norway
Fabio Oliva, Italy
Anna Osello, Italy
Alessandro Pagliaroli, Italy
Mario Patrucco, Italy
Francesco Peduto, Italy
Giorgio Piaggio, Chile
Giovanni Plizzari, Italy
Sebastiano Rampello, Italy
Jan Rohed, Norway
Jamal Rostami, USA
Henry Russell, USA
Giampiero Russo, Italy
Gabriele Scarascia Mugnozza, Italy
Claudio Scavia, Italy
Ken Schotte, Belgium
Gerard Seingre, Switzerland
Alberto Selleri, Italy
Anna Siemińska Lewandowska, Poland
Achille Sorlini, Italy
Ray Sterling, USA
Markus Thewes, Germany
Jean-François Thimus, Belgium
Paolo Tommasi, Italy
Daniele Vanni, Italy
Francesco Venza, Italy
Luca Verrucci, Italy
Mario Virano, Italy
Harald Wagner, Thailand
Bai Yun, China
Jian Zhao, Australia
Raffaele Zurlo, Italy

Strategic use of underground space for resilient cities

Tunnels and Underground Cities: Engineering and Innovation meet Archaeology,
Architecture and Art, Volume 10: Strategic use of underground
space for resilient cities – Peila, Viggiani & Celestino (Eds)
© 2020 Taylor & Francis Group, London, ISBN 978-0-367-46878-1

Designing complex urban excavations and BIM: The story of the new underground bus terminal in Slussen – Stockholm

M. Alzouby
WSP Sweden, Luleå, Sweden

ABSTRACT: To keep up with Stockholm's rapid growth, the Slussen area is being revitalized with a new bus terminal built in bedrock. Buses will circulate inside the underground terminal hall around a central waiting area, and exit at ground level on the quay-side. This requires the excavation of more than 273 000 cubic meters of blasted rock to create three large, parallel caverns. Each cavern will be up to 28 meters in span width and 200 meters long. Nearby urban infrastructure and existing rock caverns have provided major challenges to the design of the terminal. BIM comes in handy when carrying out a production-friendly design that takes into consideration the surrounding complexities and limitations. BIM and 3D models have not only became a better alternative to traditional drawings, but are also essential tools for coordinating between different technical disciplines and highlighting risk management throughout the design phase.

1 HISTORY AND BACKGROUND

The Slussen area in Stockholm has been rebuilt every 100 years since the 17th century to accommodate Stockholm's growth. Today Slussen is the second largest hub for public transport in Sweden and the transfer point between the subway, city busses and suburban busses that carry tens of thousands of travelers every day. The facility is part of a larger complex that includes bridges with service and retail stores underneath.

After more than 70 years, since the last time Slussen was renovated, today's facility is in bad condition and regular maintenance is no longer enough to keep the facility alive (Tolesson 2013). Withered concrete and rusted reinforcement in addition to foundation problems require that the facility must be demolished and rebuilt. The bus terminal is currently exposed to weather conditions and travelers must cross bus paths in order to come to their bus stop.

Rapid growth in the Stockholm region is another reason that makes rebuilding Slussen a necessity. The hub shall be adapted for the future of Stockholm. The equivalent of two full local busses of people arrive as new citizens of Stockholm every day. One bus of newborns and one bus of migrants to the city.

The reasons above show the urgent need for a radical solution in the Slussen area, which has motivated the City of Stockholm's decision to build a completely new underground bus terminal in bedrock (Katarinaberget). Stockholm City, Stockholm County and the state will jointly finance the new terminal.

The process to redevelop Slussen began with discussions as early as the 1970s. Designing and planning of the new bus terminal started in 2012 and the construction works were planned to start in 2013 and end in 2019. However, the first plan was rejected by the environmental court in 2013 due to several complexities. Only in May 2018 did the environmental court approve the new plan for Slussen (Gustafsson 2018). In late summer 2018, Implenia won the contract for construction (Time and Material) as a general contractor. The detailed design has started in early 2018 by WSP Sweden and will continue with coordination with the contractor in a so-called partnership agreement.

Figure 1.   Location of Slussen in the middle of Stockholm.

2   THE BUS TERMINAL

The bus terminal will be excavated mainly by the drill and blast method. However, some alternative methods such as wire sawing and hydraulic fracturing are utilized in some limited areas. The estimated amount of rock to be excavated is about 273 000 cubic meters. The estimated cost for the construction of the terminal is 150 million US dollars.

The entrance of the buses to the underground bus terminal is planned through a ramp from the quayside. The tunnel portal's span in the existing rock slope is ca 29 m. The terminal itself consist of 3 large caverns; one for arrivals with 6 drop off parking spaces and 20 other parking spaces used between trips; one for departure with 17 parking spaces, and a 200 m-long waiting hall between the arrival and departure caverns. The span width of these three large caverns will vary between 18 m (waiting hall) and 24 m (arrival hall). The waiting hall will be connected to the arrival and departure halls by 18 connection tunnels. To support these large caverns, the rock between the connection tunnels has been saved and designed as support pillars. These pillars have a significant importance for the overall stability of the facility.

Contrary to the norm in Sweden, traffic in the bus terminal will be left-handed. When the buses enters the terminal, they arrive in the arrival hall south of the terminal. The buses then continue, rotates and pass under the entrance hall to arrive at the departing hall. Through this short trip, the busses will be driving under several existing rock caverns with low rock cover.

2.1   Entrances and portals

The travels will be able to enter the terminal through 6 entrances leading to the eastern and western entrance halls.

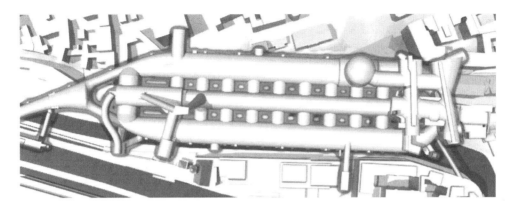

Figure 2. Overview of the terminal between and under the existing buildings.

In addition to the large bus portal, three entrances for travelers are designed from the quayside; one where the Saltsjöbana tram's entrance is today, one where the Subway entrance is today and a third one right across from the Birka ferry terminal is. The latter will be constructed as a tunnel that actually passes through a historical landmark of stairs that scale the slope. The > 100 years old stairs shall be preserved, but in order to excavate and build the entrance tunnel, they will be deconstructed, renovated and constructed again when the tunnel is finished. Every old block has been tagged and will be used when the stairs are rebuilt.

The main entrance for travelers is in an existing underground parking garage. This entrance will even be connected to the subway and Saltsjöbana tramway. Another entrance will be built in the existing subway entrance at Götgatan street. This entrance will provide a completly indoors, climate controlled, passage for subway travelers to the new bus terminal. This entrance requires sensative excavation works only a very short distance from the subway platform. The design takes into consideration that excavation and construction works during the breakthrough must be performed without a stop in subway traffic.

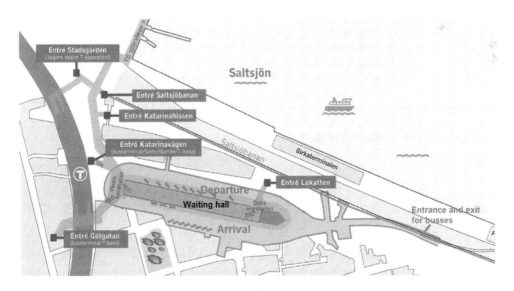

Figure 3. Overview of the terminal showing the six travelers entrances.

3 BIM, THE FIRST STEPS

In contrast to other technical disciplines in the building design and construction field, only some first attempts have been done in applying BIM in tunneling. Sweden's underground industry has been one of the leaders when it comes to applying BIM in designing tunneling and underground infrastructure projects. The new bus terminal in Slussen was the first project that WSP Sweden contracted as a wholly 3D-modell based design in 2012.

All rock excavation designs are delivered as 3D models decreasing the number of 2D drawings significantly. 3D modelling was used by almost all technical disciplines resulting in an overarching 3d project coordination model. The owner, the contractor and the designer have used this coordinating model for several purposes. During the designing process, the coordination model has been used on a regular basis in so-called VCD (Visual Communication Design) meetings.

During VCD meetings several technical disciplines sit together moving through and discussing the coordination model. The meetings are dedicated to detecting collision points and optimization needs. The outcome is a better design with coordinated solutions between disciplines.

In addition to the designed structures, the coordination model includes a rock surface model, digital maps and scanned models of existing structures and rock caverns in the area.

In a complex projects as this one where the terminal is built very close to existing facilities, it is essential to have the three-dimensional understanding provided by BIM. This understanding provides fundamental information for risk management and decision support.

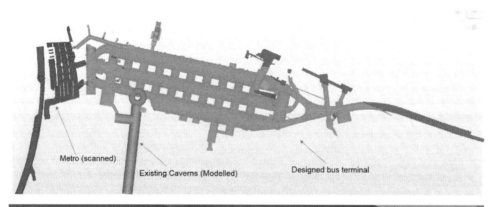

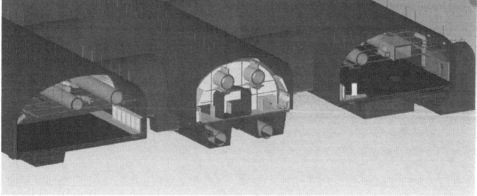

Figure 4.   View from the coordination model. Overview (top) section view (bottom).

Even though a comprehensive BIM solution is still missing for tunneling industry, these first steps have increased efficiency in this project and minimized the need for physical meeting in the VCD sessions.

3.1 Rock engineering prognosis

In Swedish system, two documents are produced by the designer which serve a similar purpose as the geotechnical baseline report: the engineering geology prognosis and rock engineering prognosis. The engineering geology prognosis is mainly presented in the preliminary design documents to the owner. This document summarizes and interpret the geological data along the planned construction and describes the ground conditions to support preliminary design. During detailed design, a rock engineering prognosis is traditionally produced as 2D drawings. In this project, the rock-engineering prognosis was delivered as a 3D Navisworks model and contains the following layers:

- Planned tunnels and caverns
- Rock Surface
- Weakness zones
- Rock quality domains
- Ground support domains
- Grouting domains
- Fire protection domains
- Other layers such as the position of coordination milestones (checkpoints) – A point were an the contractor and designer check that all is advancing as planned before further production proceeds.

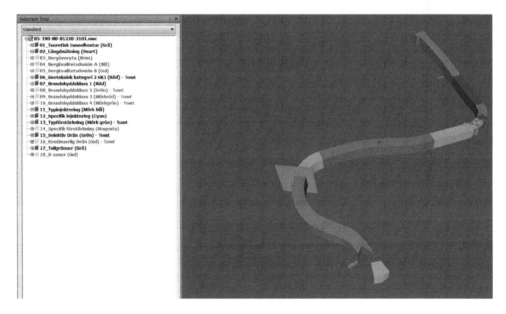

Figure 5. Rock engineering prognosis model for the access tunnel in Slussen.

4 CONCLUSIONS

The new bus terminal in Slussen is a proof of how underground facilities can be the future of our cities and urban development, especially in areas with limited surface area. Reusing abandoned underground caverns in Slussen will help the city to remain compact and relieve the pressure on the surface leaving more green and public areas. BIM in Slussen has been a very effective tool in such complex projects when visual aids are very important and critical. Dealing with third parts, property owners and authorities would not have been as simple without the visualizing aid BIM provides. BIM has been an effective tool in Identifying risks and aligning the facility parts in the most optimal location according to the nearby caverns and facities. Due to the digital visualization, the VCD meetings can be hold online allowing effective distance meetings and cuttingtime and costs of transportation for the designers. The hopes are that in the future better toolsets and faster processes can be applied over all phases of a project – development, design, construction, infrastructure management and deconstruction in the future.

REFERENCES

Alzouby, M. & Ehlis, A. 2018. Den nya bussterminalen I Slussen: Utmaningar och lösningar, Stockholm.
Alzouby, M. & Rudegran R. Modellspecifikation Berg, 2018, Slussen Berg, Stockholm.
Harer, G. Pläsken, R. Flandera, T. 2017. BIM in Tunnelling – Considerations regarding further development requirements, Sao Paulo.
Ny bussterminal I Katarinaberget, 2018. Retrieved Dec 10, 2018, from https://www.sll.se/verksamhet/kollektivtrafik/aktuella-projekt/Slussen/bussterminalen-i-katarinaberget/
Gustafsson, A. 2018. Grönt ljus för nya bussterminalen i Slussen. Retrieved Dec 10, 2018, from https://mitti.se/nyheter/gront-bussterminalen-slussen/
Tolesson, N. 2013. De får projektera Slussen. Retrieved Dec 10, 2018, from https://www.fastighetssverige.se/artikel/de-far-projektera-slussen-12426

Tunnels and Underground Cities: Engineering and Innovation meet Archaeology,
Architecture and Art, Volume 10: Strategic use of underground
space for resilient cities – Peila, Viggiani & Celestino (Eds)
© 2020 Taylor & Francis Group, London, ISBN 978-0-367-46878-1

Investing in urban underground space – maximising the social benefits

S. Bricker
British Geological Survey, Nottingham, UK

L. von der Tann
University College London, London, UK

E. Reynolds
Urben, London, UK

C. Bocci
WestonWilliamson+Partners, London, UK

P. Salak
Dr Sauer & Partners, Surbiton, Surrey, UK

ABSTRACT: With increasing pressure on space in cities, we are seeing greater development underground. Despite the multiple benefits of underground space, its social value is under-appreciated and no market for underground space utilisation exists. The result is that underground space is not planned, engineered or managed in a way to realise its potential value. This paper presents findings from a Think Deep UK initiative which explored the social value of underground space and evaluated the UK's Social Value Act which embraces social, economic and environmental benefits. It was found that the main drivers to evaluate social value for infrastructure projects are cost and risk which are intimately linked with the scheme's design life. As such, only tractable, evidence-based benefits are easily accounted for. It is suggested that social value frameworks should be flexible and incorporate qualitative measures of value across different timescales so that long-term benefits for future generations are planned.

1 INTRODUCTION

Social value is the contribution that projects and investments make to society and which results in a positive impact to people's lives. In theory social value frameworks were introduced to capture a broader measure of value, to consider not only market forces and financial performance but to evaluate the wider societal impact (both positive and negative) to demonstrate increased 'value for money' (SORI Network 2011). In practical terms there are many different definitions and interpretations of social value represented in a wide range of assessment frameworks.

The Public Services (Social Value) Act (2012) was introduced in the UK to evaluate the social value of new public services. The act requires the social, economic and environmental benefits of the public services to be considered at the commissioning (pre-development) stage in order to design better services and find innovative solutions to maximise the potential benefits (DCMS, 2018). Large government projects in the UK will often adopt a 'Benefits management' framework to fulfil this new requirement. Whilst the act only applies to public services "social value" as a concept is increasingly being applied by both public and private bodies across the infrastructure sector and wider built environment (ARUP, 2017). However, there is a perception that the rigidity of the social value assessment frameworks can inhibit the concept of social value being harnessed to enhance collective community benefits. There is also an acknowledgment that environmental considerations are not always well accounted for when social value assessments are applied in their narrowest definition. Failure to fully account for all the wider benefits within the options appraisal and business case can lead to a 'value gap' where the development doesn't deliver its potential value and the return on investment is lowered (ARUP, 2017), or an 'opportunity cost' where an alternative development or design option might have yielded enhanced benefits or reduced societal impact.

The balance between initial investment and societal benefits is particularly pertinent for underground development and subsurface utilisation, where a number of subsurface developments are necessary to support public services e.g. transport tunnels, waste facilities, public utilities. Underground development is often viewed as problematic and while the short-term costs and impact of the development are often higher than comparable development at surface, this is often outweighed by the longer-term societal benefits. In consideration of the social value of urban underground development there are a number of questions to investigate, i) Is investment in underground development preferable to development at surface? ii) If space limitations in our cities necessitates underground development, how do we maximise societal benefits?, and iii) Where the impacts of underground development are dynamic and felt across multiple temporal and spatial scales, how do we best balance individual preferences, community benefits and national interests?

This paper discusses how the social benefits of underground development may be enhanced based on a review of social value assessment frameworks, evaluation of the social value of different subsurface uses and elicitation from domain experts during a workshop facilitated by Think Deep UK. For social value frameworks to be effective for underground space developments they have to incorporate different time scales and invite stakeholder engagement early on in the development process.

2   WHY SHOULD WE VALUE THE SUBSURFACE?

There is increased pressure on underground space in our cities. Cities are growing, and with increasing pressure on space, higher land prices and a drive for compact, resource efficient cities, the use of urban underground space is broadly increasing in line with population growth (Bobylev, 2016).

The urban subsurface space is used for a wide variety of applications that deliver social benefits (Figure 1). In the UK, cities and towns have evolved to use and exploit the urban subsurface in a multitude of different ways, for example for water supply, transport infrastructure, buried utilities, and waste disposal. These different services and functions can, for example, be classified as follows (de Mulder et al., 2012):

source of natural resources
storage of materials (solid, liquid, gas)
space for public and commercial use
space for infrastructure
medium for foundation for construction
archive of historical and geological heritage

Over and above these traditional uses of urban underground space, there is increased recognition that the ground is an important component of life-support systems and delivers a range of ecosystem services, e.g. water and heat storage and conductance, such that more effective use of the subsurface can contribute to climate resilience, a low-carbon economy and sustainable living (Rawlins et al., 2015; Vermooten, 2015).

While these subsurface services and functions deliver a range of economic, environmental, social, cultural and political benefits they are not well evidenced and as such underground space remains under-valued. Strategic utilisation of underground space can be an enabler of city functions (e.g. buried utilities, water supply) but uncoordinated or fragmented planning of the subsurface can also be an inhibitor of city development and services, where subsurface

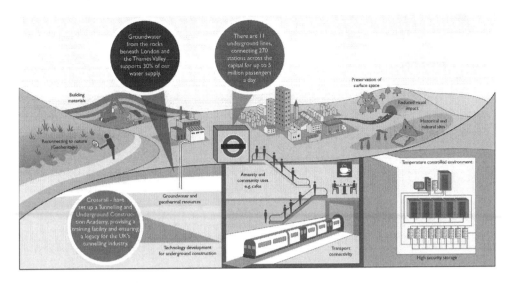

Figure 1. Illustrative example of the social value delivered by underground space.

space becomes sterilised for future uses or where subsurface uses impact negatively on other functions. To derive enhanced value from urban underground space assessment of the interactions of subsurface uses and greater coordination of underground development through the planning framework is needed.

3 MEASURING SOCIAL VALUE

To evaluate the application of social value measures to urban underground space, Think Deep UK, a group of built-environment experts, used an exploratory approach to consider how to determine the multiple benefits provided by the ground and how to measure the social value delivered by underground space. A review of social value assessment frameworks was complemented by a workshop facilitated by Think Deep UK, a group of build environment experts, in which the social value of different subsurface uses was evaluated. During the workshop, participants were asked to apply principles of social value assessments to outline the social value or benefits delivered by underground space, whether existing means of evaluating social value are adequate for underground development projects, and whether public and private works should be dealt with differently. Participants were further asked to consider both the challenges and opportunities of embedding social value in business cases for underground development and subsurface utilisation.

The varied interpretation of social value gives rise to a multitude of frameworks, guides and policy documents for social value assessments, while this provides flexibility in the application of social value it can also introduce ambiguity and inconsistency in the measurement of social value. Social value may be defined in its narrowest sense where only societal impact is considered while other definitions are broader and embrace environmental and economic benefits more akin to sustainability appraisal or Natural Capital assessment. Some frameworks encourage economic valuation and monetary indicators to measure social value such as the UK's HM Treasury Green Book Guidance (HM Treasury 2018) which is used for the procurement of government projects. The Green Book is built on the principle of a social cost-benefit analysis calculated over the design life of the development, which accounts for economic, financial social and environmental impacts, comparing options to the baseline or 'status quo' along with a five-point business case – strategic, economic, commercial, financial and management - where social value case is incorporated in the economic case.

Table 1. Stages of a social value assessment.*

Stage	Assessment
1. Total assessment of the environmental, economic and social factors	The big picture: The initial assessment considers the investment or 'input' to the project and the 'outputs' that are generated from the investment. What is changing as a result of the project? All the economic, environmental and social factors that contribute to the project are identified.
2. What are the impacts and outcomes?	Ask the right questions: It's important to look beyond the inputs and outputs: What are the outcomes from the project? An outcome is a change that occurs over the longer-term. What impact will it have? What would have happened anyway? Consider who will be affected and at what scale will the impact be felt.
3. Can these outcomes and impact be quantified?	What gets measured gets valued: An evidence base to monitor and evaluate the change that occurs as a result of development allows the outcomes and impacts to be measured, and then valued. Not all indicators of change are monetary, e.g. number of jobs, and area of land protected, are other metrics that could be used.
4. Can the options be adjusted to optimise social value?	Value is in the eye of the stakeholder: Social value considers the impact on society and people's lives. It's important to consider the priorities of the project and the stakeholders, consult with them, and identify shared priorities and potential conflicts to deliver a solution that maximises the benefits.

* Adapted from PwC 2017; London Business School 2004.

Other frameworks adopt more qualitative measures for example, the Total Impact Measurement and Management tool (PwC, 2017), the Total Value Capture tool (ARUP 2018) and the Five Capitals Model for sustainable development (ref), all of which embrace the concept of total impact measurement across financial, social, human and natural capital.

There are four main stages of a social value assessment (Table 1). These stages facilitate the identification of the investment and effects of a project, evaluation and quantification of the impacts and outcomes and finally options appraisal and optimisation to enhance the benefits and 'value for money'. Whilst these are described in a linear fashion, an iterative process that embeds public engagement at multiple stages was the preferred engagement method highlighted during the workshop, as evaluation of long term and wide-ranging societal benefits and impacts is complex and difficult to communicate. Experts commented that public consultation throughout the planning and evaluation process is proven to be highly effective, it helps to resolve conflicting priorities at an early stage, identify innovative design solutions, and helps balance functional elements with more creative, community development options.

4 EVALUATING AND COMMUNICATING THE SOCIAL VALUE OF URBAN UNDERGROUND SPACE

The review suggests that urban underground space is not routinely considered as a development option and is often viewed as an abstract concept. At the workshop, participants emphasized the perception that underground development is expensive and disruptive; this perception inhibits a more holistic evaluation of options and consequently means that potential benefits are not being realised. However, the sector has seen an increase in innovation and advanced technology to facilitate improved underground utilisation for example, boring technology for tunnel construction and aquifer storage and recovery for public water supply. If smart use of underground space is designed at the outset, as a result of more complete options appraisal which embraces new innovations, the social return on investment can be enhanced. From an urban planning perspective, greater utilisation of underground space, compared to development at surface, can provide more flexibility in urban planning and allow the surface of cities to be prioritised for higher value land uses and needs.

At present cost-benefit analysis remains the predominant tool to evaluate and compare project proposal. However cost-benefit analysis for underground development often fails to identify the broader and long-term societal benefits of subsurface utilisation. These benefits are delivered at a range of different spatial and temporal scales - the initial costs, which are often high, may be borne by a private investor but the social value that is delivered by underground space utilisation is often greater than the capital expenditure, and it is delivered over longer timescales and at multiple spatial scales. In such cases, a cost-benefit analysis based on financial metrics will conclude a low financial return on investment, since broader social value is neglected. Given that underground development occurs at the intersection of the natural (the ground), built (physical infrastructure) and social (public services) sphere, a broader social value assessment framework which captures the social, environmental and economic impacts is expected to be more effective.

A more sophisticated life cycle and cost benefit analysis will allow the intrinsic values, environmental services, and competing demands on underground space and resources to be evaluated and the 'value gap' and 'opportunity cost' to be reduced. This requires a robust evidence base including information about existing subsurface utilisation, what functions might be displaced or impacted as well as visions for future uses. Underground space is finite resource and should be managed effectively. Development of the evidence base would support better informed planning policy and more effective selection of development or use options.

Despite the opportunities in social value assessment of underground development, a number of complexities were highlighted by industry experts, which can act as a barrier to its implementation. There are multiple organisations and stakeholders, both public and private, that currently use and derive value from the urban subsurface space. For the social value of underground development to be fully embedded and mechanism to balance the tradeoffs between individual gains, community benefits and national interests needs to be in place.

Whilst a clear definition of the terms of social value and what this entails is needed, experts considered it more important to discuss the societal impact early in the project development and identify the 'value gap' and potentially missed opportunities. Individual successes where public consultation had led to enhancement of social value – with associated financial savings – were highlighted and early stage consultation with potential beneficiaries and community-led engagement are considered crucial for this success. With this type of approach, the full potential of underground development, which may have a higher initial cost but greater long-term benefits, could be realised.

Wider stakeholder engagement and the promotion of public-private partnerships would also facilitate 'impact investment' for underground development. Impact investment is a growing market, worth £150m, where investors and fund managers are committed to increasing the social return, as well as financial return on investment in the UK (ARUP 2018). Social impact investment products and reporting tools for non-financial outcomes are being developed to support this market. Adopting a broader framework to measure the social value of underground development can increase the opportunities to identify new investors and funding mechanisms, i.e. the social impact is more visible and those that benefit are more motivated to invest. Understanding the integrated value supply chain for underground development and subsurface utilisation is therefore critical.

5 RETHINKING SOCIAL VALUE FRAMEWORKS FOR UNDERGROUND DEVELOPMENT

Evaluation tools like CBA start to consider social value and wider society benefits but usually these assessment forms part of the business case and the according processes favour quantitative measures, are often domain-specific with projects considered in isolation and cost and risk for the specific domain are primary drivers for the evaluation process. As such, only tractable, evidence-based social benefits can be easily accounted for. The more qualitative impacts and considerations are difficult to capture in this style of assessment and framing a suitable process for social value assessment is challenging as different benefits occur at a multitude of levels, and may take a long time to be fully realised. Therefore, project development needs to be refocused from a purely economic endeavour to an inclusive process that embraces socio-economic indicators and appropriately weighs project proposals against other potential uses.

Communicating the social benefits and impacts is key to explaining the value of underground space utilisation and making the benefits more visible. Discussions about social values, benefits or impacts of underground activities cover a wide range of topics and will differ by project, location, stakeholders and cultural setting. Context specific settings have to be considered as social value is often discussed in a particular and unique environment and people likely to be affected rather than with regards to a wider national or, potentially global discussion. Engaging in national and local politics, communicating with key stakeholders and the public, through multi-disciplinary approaches is therefore critical to a successful outcome.

In summary, to the apply the concept of social value in general and regulations like the Social Value Act effectively for underground space the following recommendations are made:

- The social benefits of underground space utilisation need to be better understood and articulated.
- An evidence base to help measure and evaluate the benefits and impacts of underground development needs to be developed. Identification of the integrated value supply chain for subsurface utilisation would support this.
- Those who commission and undertake assessments need to know the benefits that use of underground space brings, to consult potential beneficiaries and affected communities early on and to have the means to include the value of those benefits within their assessment.
- The benefits and limitations of underground space utilisation need to be considered fully at the pre-commissioning stage and in parallel with planning policy. Exploring governance options for underground space, such as 'public commons' use may help facilitate shared use of underground space and help protect high-value uses such as public groundwater supply and transport networks.
- A framework to assess social value should be developed that is flexible enough to incorporate qualitative measures of value, across different timescales such that long-term benefits and broader societal needs of future generations are planned for. Application of impact investment products for underground development should be explored.

6 CONCLUSION

This paper showed that integration of social value in cost-benefit analysis is enshrined in the Public Services (Social Value) Act and also embedded in number of other value frameworks. However, the applicability of these frameworks to assessing underground developments and utilisations is not straight forward and only tractable, evidence-based benefits can be easily accounted for in commonly applied tools like cost benefit analysis. To facilitate social value assessments for underground space uses, the benefits of underground development need to be better defined and the opportunities clearly articulated to decision-makers. A broad evidence base, assessment skills and flexible assessment tools that integrate long-term considerations are suggested as major aspects to be developed to enable efficient integration of social value in project appraisal schemes and a change of focus from purely economic endeavour to a more socially inclusive process.

REFERENCES

ARUP 2018. Making the Value Case for Infrastructure. Unpublished Report.

Bobylev, N. 2016. Transitions to a high density urban underground space. 15th International scientific conference "Underground Urbanisation as a Prerequisite for Sustainable Development". *Procedia Engineering*, 165: 184–192. doi: 10.1016/j.proeng.2016.11.750.

DCMS 2018. Department for Digital, Culture, Media and Sport. The Public Services.

Social Value Act 2012. An introductory guide for commissioners and policymakers. https://assets.publishing.service.gov.uk/government/uploads/system/uploads/attachment_data/file/690780/Commissioner_Guidance_V3.8.pdf (accessed 06.09.18).

HM Treasury 2018. The Green Book. Central Government Guidance on Appraisal and Evaluation. ISBN 978-1-912225-57-6.

de Mulder, E., Hack, H. R., & Van Ree, C. C. 2012. Sustainable Development and Management of the Shallow Subsurface, Geological Society; London. ISBN 978-1-86239-343-1.

SORI Network 2011. Commissioning for Value Thinking, Think Piece. http://www.socialvalueuk.org/app/uploads/2016/03/commissioning%20for%20value%20thinking.pdf (accessed 06.09.18).

PwC 2017. Total Impact Measurement and Management Framework. https://www.pwc.com/gx/en/services/sustainability/total-impact-measurement-management.html (accessed 10.09.18).

Rawlins, B. G., Harris, J., Price, S., & Bartlett, M. 2015. A Review of Climate Change Impacts on Urban Soil Functions with Examples and Policy Insights from England, UK. *Soil Use and Management*, 31: 46–61.

Vermooten, S. & Lijzen, J.P. 2015. Ecosystem services of the groundwater and subsurface. RIVM report 2014-0167 https://www.rivm.nl/en/Documents_and_publications/Scientific/Reports/2015/juni/Ecosystem_services_of_the_groundwater_and_subsurface (accessed 10.09.18).

Tunnels and Underground Cities: Engineering and Innovation meet Archaeology,
Architecture and Art, Volume 10: Strategic use of underground
space for resilient cities – Peila, Viggiani & Celestino (Eds)
© 2020 Taylor & Francis Group, London, ISBN 978-0-367-46878-1

The construction of the underground car park beneath the "San Giusto" hill, in a very delicate urban environment in Trieste, Italy

M. Bruga, A. Bellone & M.A. Piangatelli
CIPA S.p.A., Rome, Italy

ABSTRACT: The city of Trieste, like many urban centers, suffers from a lack of facilities dedicated to car parking. Thus, in 1996, the idea of creating a parking space under the San Giusto hill was born. CIPA S.p.A. carried out the works related with the tunneling (consisting in the execution of the excavation works and the lining of the ventilation shaft, the entrance tunnel, the caverns, the transects, the pedestrian tunnel and other minor tunnels), as well as those related with the construction of all the reinforced concrete structures. The caverns were approximately 120 m long with a total section of approx 315 m2, inside which the multi-storey buildings have been built. The ventilation shaft had a depth of 63 m and an excavation diameter of 9.60 m. It has been executed with the underpinning methodology, with 2 meters each step. The location of the work and the characteristics of the tunnels turn this work a fascinating one but at the same time delicate enough to require attention and particular technical expertise during its execution.

1 INTRODUCTION

The idea of creating a parking space under the San Giusto hill started in 1996 with a first technical proposal to the "Pubblica Amministrazione di Trieste". After a study of feasibility and following a formal proposal for the implementation of the Project Financing, which took place in 2002, after the planned tender procedure was completed, in 2005, the Concession for the Design, Construction and Management of the parking area was given to the "Proponent" (today Park San Giusto SpA).

Since then, and due to geotechnical, environmental, urban and archaeological constraints, the "Project" went through substantial changes up to the final design version (2009) and subsequently, in 2011, to the final version of the Execution Project. Park San Giusto SpA proceeded with the internal assignment of the works to the construction companies, initially united in a Temporary Joint Venture with RICCESI SpA, CARENA SpA, C.E.L.S.A. Soc. Coop., Arm Engineering SpA and Mecasol S.r.l., which subsequently integrated the consortium PIESSEGI S.c.r.l. The consortium then entrusted CIPA S.p.A. the works related with the Tunnelling and the construction of all the internal reinforced concrete structures.

2 STRUCTURE OF THE PROJECT

2.1 Functional and architectural description – The parking structure

The underground parking is located under the area partly occupied by the "Seminary" and the "Convitto delle Monache" on the slopes of San Giusto's hill. The parking structure is completely underground, and consists of two caverns with 5 parking levels, 17x120x18m each (Figure 1). The caverns are connected by two transepts for vehicles and by tunnels towards the two exits/accesses located one in Via del Teatro Romano (for vehicles and pedestrians) and one on the Colle di San Giusto (for pedestrians) served by two elevators (Figure 2). Inside the two caverns there are two one-way ramps connected by a two-way transept that connects the five levels which give a capacity of 718 parking spaces in a total area equal or less than 1,500 m^2. One of the parking levels (level 0) is placed 3.0 m above the street "Via del Teatro Romano" and another one (level +1) is 9.40 m above it.

The other three levels are placed below in steps of -2,60m. All five levels are two to two intercommunicating through a two-way front transceiver; the three "lower" levels, intended for rotation, are also connected by a rear service transept, in addition to the front one, for an easy ring circulation. The internal system has 5.50 m wide lanes, with a single anti-clockwise direction, which is in compliance also with the rules also accept for circulation in two directions. The connecting transects are 8.20 m wide. The one-way ramps have an external radius of 8.00 m and lanes 4.50 m wide.

The pedestrian exits are served by four escalators, two of which with elevators leading to a pedestrian path which, in part, also acts as a safe place in case of fire. Access and exit of vehicles is planned open-air and part in an artificial tunnel directly from the Roman Theater.

The parking is connected with two lifts to the Colle di San Giusto, an important tourist destination in the city. The pedestrian accesses are double; the main one from Via del Teatro Romano follows an isolated route from the vehicular one along the entrance gallery; the second access is placed, opposite, in the back and leads directly to the top of the Colle di San Giusto for direct access to the archaeological site through two elevators.

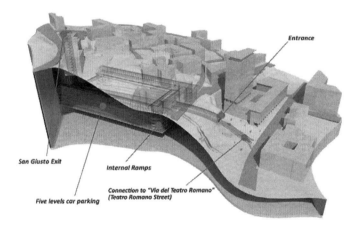

Figure 1.   General view of the parking. The entrance for vehicles in the street "Via del Teatro Romano" and the entrance for pedestrian in San Giusto.

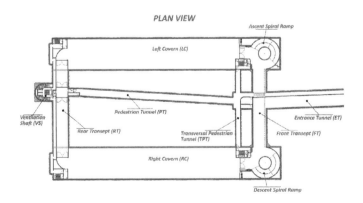

Figure 2.   General plan of the parking Project.

2.2 Geological - geotechnical general plan

The area studied is located in the city centre of Trieste, at *Colle di S. Giusto*, between *Via del Collegio, Androna San Saverio, Via delle Monache, Piazza San Cipriano* and *Via della Cattedrale*. The "territory" involved is placed at the passage between the "embryonic structure of the Ciceria" (zone of vicarious faults of the "Linea di Palmanova") and the "Thrust di Koper" which follows to the South.

Most part the area is affected by the outcrop of stratified clastic sedimentary rocks, all referable to the formation of the flysch. It is a formation of Eocene age made up of regularly varying marls and sandstones: the sandstones are normally of variable thickness from a few centimetres to a meter, the marls from a few millimetres to 40-50 centimetres. In the province of Trieste this rhythmic repetition of sandstone and marl is variable in the different localities both for the individual power of the two lithotypes as for the percentage ratio of one lithotype to the other. This is also found in areas not very extensive and is often accompanied by significant phenomena of tectonization. The rock mass as a whole is to be considered semipermeable, with discrete physical and mechanical characteristics even though, given the petrographic, physical, chemical and hydro genetic characteristics it can be present in particularly tectonized volumes (narrow folds, fault planes, interlaying slips or folds-faults).

The structural arrangement is quite complex, even if its complexity results more from the overall plasticity of the flyschoid formation than from established tectonic lines. The main tectonic motif in the area consists of a wide radius bend (with weak curvature) with NW-SE Dinaric Alps axis with immersion in this direction, while the local structure referred to the San Giusto hill is affected by a series of vicarious faults, also with the orientation of the Dinaric Alps.

The formation investigated in depth, -50 meters from the surface and over a large area, shows a very variable configuration and characteristics. From outcrops with powerful thicknesses of prevalent banks of tenacious sandstone alternating with subtle levels of low stressed marls, we pass to mainly strongly tectonized marly levels.

2.3 Principles of the structural design and planned work phases

The project foresees a widespread primary consolidation of the excavation volumes with rock riveting and active tie rods linked to a shotcrete coating reinforced even with lattice girder steel ribs. The primary consolidation includes all the elements installed in the final profile to achieve the stability of the excavation during the construction period, such as shotcrete and ribs, steel mesh, anchors in rock etc. All elements of primary consolidation must remain in place and become part of the definitive works. For the vaults of the caverns there is a further structure in reinforced concrete with load-bearing function. At the end of the excavations and of primary consolidation, the internal structures of the car park are planned (car spaces, final cladding, floors, ramps, etc.).

The consolidation class is calculated according to Austrian standards ÖNORM B 2203-1. The determination of the consolidation class must be established on site, due to the local conditions found during the excavation. A specific consolidation class is represented by two numbers. The first reference number depends on the maximum digging length allowed before the primary consolidation is installed, and the second reference number depends on the quantity of the primary consolidation elements to be installed relative to a digging length equal to one meter.

Since these are works in an urban environment, it was not possible to use explosives during the excavation phases, and so the works proceeded exclusively with mechanical systems by demolition hammer and/or with a precise drill. The complete progress cycle includes: the excavation for a length compatible with the local geological conditions and with the consolidation class foreseen in the project; the photographic and descriptive relief of the face; the laying of the 5 cm thickness pre-shotcrete; the laying of the first layer of electro-welded mesh (ensuring an overlapping of at least 20 cm, both transversely and longitudinally); the laying of the rib; the laying of the first phase shotcrete with 17 cm thickness; drilling, laying and injection of the bolts required by the project; installation of the plate and nuts at each bolts; the laying of the second layer of electro-welded mesh, (always guaranteeing the overlap); the laying of the finishing or second phase 8 cm thick shotcrete, all in a covering package with a total 30 cm thickness.

3 EXECUTION OF WORKS AND ISSUES FOUND

3.1 *Construction site set up*

For the execution of the works, two distinct areas were identified corresponding to the two work sites:

- site area "A", in *Via del Teatro Romano* at +5.00 m asl: excavations and lining of the tunnel entrance, caverns, transepts and the pedestrian tunnel were carried out (Figure 3);
- site area "B", in *Via Rota* at an +60.60 m asl: excavation and the lining of the ventilation shaft are to be carried out (Figure 4).

In the site area A there were: the offices, the lockers rooms and toilets for the workers, the electrical and mechanical workshop, the temporary storage of the excavation material and the infrastructures for the power supply, electric, ventilation station and tanks for the intake and fuel accelerator. In the same space some small areas were used for the storage of various materials and equipment also necessary for the execution of the works. In area B a tower crane was installed to serve the shaft; the limited residual space was used as a temporary storage area for the materials before they were transferred to area A.

One of the most challenging issue during the execution of the works, especially in area A, was the coordination of the traffic flows within the construction site area; several times in particular situations of congestion we have been forced to slow down or to interrupt and/or postpone entire work phases such as the execution of an excavation or a concrete casting.

Due to a total lack of space it was not possible to install an automatic system for washing the concrete truck mixers; this led to a daily need to submit the trucks to a long and costly manual washing at the end of each single unload operation. Furthermore, since it was not possible transporting of the excavated material from the excavation face directly to the final destination (the reduced section of the caverns did not allow the access of some vehicles), it was necessary to create an area for temporary deposit that occupies a large portion of the site establishment (the capacity, however, had to be such as to be able to deposit the material dug during the weekend when it was not allowed the truck transit).

The limited capacity of the storage areas also required a "just in time" management and procurement of construction materials: at the site could be downloaded and stored only the materials to be installed/used in the immediately following hours (ribs, electro-welded mesh, steel bars for reinforcement etc...).

All machines and equipment had to be parked inside the tunnels and also all ordinary and emergency maintenance interventions had to be performed underground in very small spaces.

The situations above mentioned represented a strong limitation every day for the continuity of the work and required to the technical staff a large organizational effort in order to limit as much as possible the interventions.

Figure 3. The site area A in Via del Teatro Romano (in the background the temporary storage of the excavated material).

Figure 4. The site area B in Via Rota during the excavation works of the shaft.

3.2 Excavation and lining of the ventilation shaft

The Ventilation Shaft (VS) connects the rear transept (RT) with the surface on *Colle di San Giusto*, between *Via del Castello* and *Via Rota* and will accommodate many services including stairs, two lifts, ventilation and air extraction ducts. It is approximately 63 m deep with an excavation diameter of 9.60 m. It was done by an underpinning methodology, in steps of 2 m (Figure 5 and Figure 6). The excavation of each single section of the shaft was carried out with a 1.2 ton crawler excavator with a demolition hammer and a simultaneous extraction of the excavated material using a self-unloading bucket transported by a tower crane mounted at the surface.

After each excavation step and the subsequent laying of a draining membrane along the perimeter, the reinforcement was laid down and installed and then the circular formwork, so that the concrete casting could then serve as final lining. After the removal of the formwork, the excavation for the next segment started. At this stage the starter bars from the previous step that were previously involved in sand, were discovered. The VS work took 8 months to completion, with excavation and lining working 24 hours/day. The particularly favourable conditions of the rock mass allowed to proceed directly with the execution of the final lining without requiring any temporary support (shotcrete, ribs or anchors).

3.3 Excavation of the tunnel entrance and of the front transept

The Entrance Tunnel (ET, Figure 2) starts from the portal in *Via del Teatro Romano*, arrives at the front transept (FT), and is approx 42 m long, with an excavation section of 47 m^2 and an average slope of 2.2%; it is used by vehicles and pedestrians and is also used for the ventilation. The Front Transept (FT) connects all five levels of the right and left caverns; in addition the upper part is used for the ventilation. It is approx 74 m long (of which 9 m intersecting with the ET), for a total section of approx 135 m^2 excavated in successive phases: a first excavation section of 48 m^2, and the subsequent ones in five different phases.

The ET excavation was full face, with a consolidation class with steps of 1.00 m, laying of ribs, consolidation of the rock with passive anchors connected to a shotcrete lining reinforced with electro-welded mesh and carried out in three successive phases. The ET excavation required 65 days of work in two shifts (from 6.00 to 22.00) with an average daily production of 0.65 m. The biggest limitation for the productivity were the allowed working hours; in order to reduce the noise impact on the surroundings' population, it was impossible to work during the night near the ET. Another issue was the interference of an existing tunnel (used as a refuge area during the war period) with the path of the ET (and with the FT and the caverns), which required previous works of filling with concrete and soil; the presence of the tunnel has also greatly slowed down the injection operations (Figure 7 and Figure 8). After the excavation of the ET, the work proceeded with the excavation of the FT which took place along two fronts.

The consolidation class adopted was with full-section with steps of 1.30 m, rib laying, drilling and laying of 4/5 passive bolts in bars Ø25mm with a length of 6 m in the shell, drilling and laying of 4 passive anchors in bars Ø32mm with a length of 12 m on the piers and lining with 30 cm of shotcrete reinforced with electro-welded mesh and performed in 3 phases (pre-shotcrete before laying the rib, first stage shotcrete before the drilling of the second phase anchors and final shotcrete layer).

In addition, at the intersection of the FT with the caverns, a pre-consolidation of the front was carried out with GRP bars 8 m long. The excavation of the FT required 60 days, working 24 h/day in 3 shifts with an average daily production of 1.10 m. The progress towards the interior allowed an increase in the shifts, even though the night time restrictions remained (from 22.00 to 06.00) for activities with greater noise impact.

The equipment for the excavation and lining of the ET and the FT was a 24 ton crawler excavator equipped with a hammer, a wheel loader and two mini dumper for the slurry, a shotcrete pump, a single arm jumbo for the drilling, an injection plant for the cement mixture, a lifter with a clamp for laying ribs, hook and basket for the execution of all works at height.

Figure 5.   Shaft reinforcement and formwork.

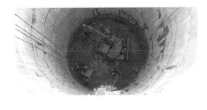

Figure 6.   The shaft excavation.

Figure 7.   Interference of an existing tunnel and a shaft during the excavation of the ET.

Figure 8.   Overview of the ET.

3.4   *Excavation and lining of the caverns (arches)*

The two caverns, right (RC) and left (LC), are the essential volume of the park structure within which five parking levels have been created, connected to each other by circular ramps, located in the front part of the caverns. The caverns have a length of about 120 m, a width of 17 m and a height of 18 m, for a total volume of approx 75.000 m^3, excavated in successive phases: a first excavation of the shell of approx 98 m^2, and then more six subsequent phases. Once the excavation of the FT finished, before proceeding with excavation of the caverns, a group of supplementary bolts was performed in the sections in the intersection with the transept constituted by 96 bolts ø32mm 8 m long for the LC and 76 bolts for the RC. This allowed the cutting of the ribs of the FT in complete safety. The excavation of the two caverns was performed simultaneously and required 152 days for the LC (0.80 m/day) and 163 days for the RC (0.75 m/day).

The typical consolidation section used in the LC was with variable steps from 1.0 m to 1.3 m and primary consolidation consisting of a lattice girder steel ribs, a 30 cm thick shotcrete package reinforced with electro-welded mesh and a radial pattern of 12/13 bolts ø25mm 8 m long. On the RC the interception in the initial stretch of a rock mass with poor geotechnical characteristics (tectonized volume due to the presence of a fault), combined with the occurrence of important surface deformation phenomena (subsidies), required the adoption for the first 40/50 m of excavation, of a more precautionary consolidation section with steps ranging from 50 to 100 cm and a radial pattern of 16/17 bolts ø25mm 12 m long.

The need to proceed with the excavation in both caverns at the same time, contrasted with the difficult environmental context characterized by the limitation of space and working hours (prohibition of noisy work, such as excavation and drilling, from 22.00 to 6.00). The most critical coordination procedure was certainly the one concerning the excavation and transport of the excavated material (Figure 9), because the only possible path for the transport of the material was the ET (Figure 10) and because of the limited capacity of the temporary storage. It was impossible working in the RC and LC at the same time. For this reason we were forced to carry out excavation shifts alternating between one cavern and the other with the consequent need to perform within two hours, two excavations and two groups of bolts with the same workers. During the eight hours of the night time, all the other operations necessary for the execution of the primary lining were carried out (laying of the electro-welded mesh, ribs, shotcrete and the injection of the anchors).

It is interesting to highlight how the limitations to the maximum length of each step have not always been determined by the geomechanical characteristics of the soil (especially in LC and in the final steps of the RC), but often by the need to close the excavation and drilling cycle on both caverns within the 16 daily working hours.

Although it was not possible to overlap the excavation work in the two caverns, the two work fronts were equipped with autonomous machines and equipment, in order to limit the difficult and expensive movements of the machines from one cavern to the other. Each work front was then equipped with a 35 tons crawler excavator with hammer, a wheel loader and two mini dumpers for the slurry, a shotcrete pump, a double-arm jumbo, an injection plant for the cement mixture and a lifter with a clamp for laying ribs, hook and basket for the work at height.

Once the excavations were completed, the project foreseen the final lining of the shell with a reinforced concrete structure made by two structural beams and an arch, before any other work.

The beams have a rectangular section of 1.20 m x 1.30 m and have been carried out after installation of pre-assembled reinforcement cages. Before proceeding with the casting the beams were anchored along the entire development of the cavern with active anchors with 6 strands placed at a distance of 1.50 m, 18 m long (in the casting of the beams, blockouts were left for later installation of the anchoring plates).

Once tie rods work was completed, the waterproofing system, a 500 g/m^2 geotextyle with a 2 mm thick layer of PVC, was installed. Along the extrados of the beam in contact with the primary lining, a micro-perforated PVC drainage pipe with 150 mm diameter was placed embedded in a layer of gravel that conveys the water to a recessed well in the beam.

Anchors equipped with eyebolts were fixed to the primary lining to support the reinforcement of the arch. This reinforcement consisted of a double layer ø16//20, crossed with a layer ø20//20. The extrados layer was fixed to the anchors provided with eyebolt while the intrados layer was suspended from the extrados one by links with double hooks.

For the installation of the waterproofing system and of the reinforcement a movable scaffolding 9 m long was mounted on rails. For the execution of the castings, a single tunnel formwork was assembled with a supporting structure in reticular steel and a 12 m long wooden mantle, initially dedicated to the execution of the castings in the left cavern and later disassembled and reassembled in the right cavern.

For the installation of the waterproofing system and of the reinforcement a movable scaffolding 9 m long was mounted on rails. For the execution of the castings, a single tunnel

Figure 9. Removing the excavated material in the right cavern.

Figure 10. Connection between entrance tunnel, pedestrian tunnel and left front transept.

Figure 11. Formwork for the arch of the caverns.

Figure 12. Final lining of the left cavern shell.

formwork was assembled with a supporting structure in reticular steel and a 12 m long wooden mantle, initially dedicated to the execution of the castings in the left cavern and later disassembled and reassembled in the right cavern. The execution of the final lining of the shell for both caverns (Figure 11 and Figure 12) took a total of 127 calendar days including the stops for the assembly/disassembly of the scaffolding and the formwork, a period during which all the other activities in the cavern had to be interrupted.

3.5   Excavation of the pedestrian tunnel

The pedestrian tunnel (PT) connects the FT on level "0" to the PT and the VS on level "-3". In addition, a cross-passage tunnel (Transversal Pedestrian Tunnel, TPT) provides the access of the right and left caverns to level "0". The PEDT is 100 meters long and has an excavation section of 25 m^2 while the TT is 42 meters long and has an excavation section of 13 m^2. The PEDT has a triple function (pedestrians entrance, safe place in case of fire and connection between the centre of Trieste and the hill of San Giusto through the VS) and during the work it was the only access road to the bottom of the excavation (caverns and transepts) when the ramps inside the front transept were eliminated.

Even though its function was relegated to the terminal phase of the underground excavations, its excavation was anticipated, overlapping it with the execution of the linings for two reasons: during the months in which the linings were carried out the transit of vehicles inside the tunnels and in general throughout the construction site area was very limited (and concentrated only on the days when the concrete castings were done), consequently the removal of the PT excavation material was much easier. The characteristics of the route of this tunnel (sub-parallel to that of the caverns and intersecting all levels from 0 to -3) turned it into a kind of pilot tunnel whose execution in advance allowed to provide important geotechnical information that confirmed the design hypotheses. The progress of the PT was done with steps ranging between 1.00 m and 1.70 m, with consolidation of the rock performed by radial bolts (8/9 bolts Ø25 3.5 m long) linked to a 30 cm thick shotcrete lining reinforced with electro-welded mesh.

Although the design did not provide any structural rib for the consolidation class in use, the contractor considered appropriate to proceed with the excavations by placing every 2 step an IPE160 rib as a jig in order to limit extra-excavation and contain the consumption of shot-crete. The excavation of the PEDT took 56 days of non-consecutive works in only two shifts (from 6.00 to 22.00) with an average daily production of 1.80 m.

The equipment for the excavation and lining of the PEDT consisted of a 13.5 ton crawler excavator equipped with a demolition hammer, a tracked loader and two mini dumpers for the excavated material, a shotcrete pump, a single-arm jumbo, an injection station for the cement mixture, a lifter for the execution of all works at height. The major difficulties encoun-tered were due to the limited space (especially the width between the two piers of the tunnel) that have greatly limited the performance of the work by obliging to perform it in one single succession without the possibility of any overlap; in particular, the excavated material could only be transported outside by suspending the excavation works and required the need for complicated manoeuvres in reverse of the tracked loader to the entrance where the mini dumper was loaded (Figure 13 and Figure 14).

Figure 13.   The confined space in the PT.   Figure 14.   The Pedestrian Tunnel excavation.

For the same reasons, particular attention was paid to the organization and verification of the procedures considering the workers' safety: during the excavation phase, access to the tunnel to vehicles and workers was completely forbidden to workers and vehicles to avoid the risk of crushing as also during the shotcrete execution phase due to the dust produced by the process that rose from the working front towards the entrance.

3.6 Excavation of the caverns and transects

Once the lining of the arches was completed, the downward excavations started. Starting from level +9.40 (excavation level) the project involved the execution of a total of 6 steps (the first with a height of 2.50 m, the next four with a height of 2.0 m and the last one with the remaining height of 2.90 m) until reaching the excavation base level at -4.00 m asl (Figure 15 - 16).

The excavation started in the **LC** (the first one to be covered) and gradually was extended to the FT and the RC. Between the first and the second levels, the parking connections to the external ventilation system were also excavated. Once the level 3 lowering excavation in the **LC** was completed, before proceeding with the subsequent level, the PT has been excavated (which connects the two caverns, the ventilation shaft and the PEDT at a lower level than that of the 5.50 m).

As the depth of the excavation was lowered, the connecting ramps between the FT and the caverns had to be reshaped. When the excavation depth has been reached in the PT, the two diaphragms that connect the transept with the PEDT on one side and the VS on the other have been cut down.

For the excavation and the primary lining execution of each level, a phasing was adopted with the dual purpose of limiting the time between the execution of the excavation and the primary lining and of distributing in time as much as possible the volumes of excavated material thus avoiding the overload of the external temporary storage.

Initially it was excavated in each cavern and on each level a volume equal to 77% of the total volume leaving the two lateral piers in place.

As soon as there was enough space for the subsequent works, the lateral piers were excavated following a "quincunx" pattern: before the excavation of the adjacent piers, the primary lining of the previous ones was completed with passive bolts (made of solid steel bars ø32 12 m long on a 1.00 m x 2.00 m grid) connected to each other with a 30 cm thick shotcrete package made in three successive phases and reinforced with a double electro-welded. Finally, before the excavation of the level n+1, the walls consolidation at level n was completed with the execution of 6 strands active anchors long between 9 to 18 m, anchored to the wall by means of a distribution plate in contact with a steel cage previously embedded in the shotcrete.

The excavation of both caverns lasted about 200 calendar days, working in 3 shifts, 24 hours a day, 7 days a week. Also for this excavation only works with reduced noise impact could be performed during the night time (injections, laying of electro-welded mesh and execution of the shotcrete).

The equipment for excavation and lining for each one of the two caves consisted of a 39 ton crawler excavator equipped with a demolition hammer, a 24 ton crawler excavator tons with a bucket, a shovel and three mini dumper for the excavated material, a shotcrete pump, a three-arm jumbo, a rig for the drilling of the tie rods, an injection plant of the cement mixture and a lifter with hook and basket for the execution of all work at height.

Figure 15.   A scheme of the excavation steps of the caverns and front transept.

Figure 16. Caverns at different excavation levels.

Figure 17. Construction of the reinforced concrete structures in the caverns.

3.7 *Execution of the internal reinforced concrete structures*

The lining of the caverns and transepts were made starting from the base and to raised up to the height of the arches previously done. The perimeter walls (25 cm thick)were executed for each level with the partition walls for the stairways, the technical rooms and the ramps, and the floor slabs (partly full-section in reinforced concrete 25 cm thick and partly in a light section with a thickness of 60 cm with "igloo" elements). A unique feature of the Park San Giusto (compared to the existing structures in the city of Trieste) is the adoption of single span slabs with few reinforcement. This solution allows having large areas without columns (Figure 17).

Mixed steel-wood formwork was used for the horizontal and vertical cavern structures. Pre-slabs slabs were used as lost formwork for the spiral ramps between the slabs. The formwork of the arches of the transepts, of the pedestrian tunnel and of the entrance tunnel was instead done using steel ribs with wooden panels; finally, a sliding formwork was used for the execution of the partitions (stairwells and elevator) inside the VS.

All the castings were carried out by concrete pipes, with the concrete mixer trucks outside, in the two site areas; so the resources necessary for the castings could be reduced inside the tunnels (consequently reducing the lost times for manoeuvres and changes), and reducing the overloads on the already finished floors and slabs, allowing an accelerated removal of temporary supports.

4 CONCLUSIONS

The large underground parkings in urban centres are a functional solution to the chronic traffic and parking problems affecting large cities. With this in mind, Park San Giusto represents a sort of "pilot" project in Italy, in addition to having represented a challenge from a purely technical and executive point of view.

REFERENCES

Sergas, F. 2011 – Park San Giusto. Un parcheggio in caverna in un centro storico con vincolo archeologico – aspetti finanziari, concessori, espropriativi e tecnici costruttivi. *GALLERIE e GRANDI OPERE SOTTERRANEE n. 100, Dicembre 2011*, 59-74, PATRON EDITORE
Grimaldi, G. 2013 –Relazione Finale Progetto Formativo: Realizzazione di un parcheggio multipiano interrato nel Colle di San Giusto, Trieste

Tunnels and Underground Cities: Engineering and Innovation meet Archaeology,
Architecture and Art, Volume 10: Strategic use of underground
space for resilient cities – Peila, Viggiani & Celestino (Eds)
© 2020 Taylor & Francis Group, London, ISBN 978-0-367-46878-1

Milan metro 4 extension to new HSR gate station "Porta Est": Underground choice and suburban regeneration

A. Bruschi
MMspa, Milan, Italy

ABSTRACT: Milan, Italy's economic capital and larger metro area, is keeping on its booming period. Boom is mostly based on "TODs" Transit Oriented Developments planned to ease and increase their accessibility, on public transport in order to enhance sustainable mobility. It's the case of "Porta Est" HSR gate station. Foreseen slightly out of Milan municipal borders in the first metropolitan ring city of Segrate, Porta Est (East Gate) will be built on Milan – Venice rail axis, thanks to aside already under construction, huge and high brand Westfield Milan mall, largest in Europe and currently – standing at about 2bn€ - major foreign investment in Italy. LIN and corresponding M4 have to be connected to gate station. The choice of extending M4 underground and TBM excavated, shifting its eastern terminus from LIN to Porta Est will maximize benefits overground allowing suburban regeneration.MM took care of a specific feasibility study.

1   MILAN "PORTA EST" FEASIBILITY STUDY

Study's purpose was to determine the evolution of mass public transport systems in support of new *Westfield Milan* Mall, under construction in a key position in the inner east ring of Milan Metropolitan Area.

Such an asset regards substantially:

○ New Porta Est HSR gate railway station;
○ Link between Westfield Milan/HSR station and LIN airport/M4 metro line

This together with the development and rationalization of the LPT (local public transport) network.

Evaluations are also made on the implementation of the road network and on complementary actions to strengthen the inter-modal role of the new hub, such as cycle and pedestrian access and interchange in general.

Study has been based on functional criteria and cost-benefit analyses (CBA), representing the ideal transport scheme to serve and interface with the territory, giving the best response to emerging needs as a result of forecast and planned developments. In practice, it's the network scheme, offer and operating program to be determined comparing alternative scenarios through a rational comparison of alternatives, technical and typological.

Giving as granted HSR gate station, necessary keystone of future local network, Westfield Milan/HSR station - LIN/M4 link has been valued in four modal options, namely:

• Bus/BRT, at grade on reserved path;
• Cable car, ropeway kind;
• Monorail, on viaduct;
• M4 metro extension, underground TBM excavated.

Solutions have been assessed basing on their capacity to respond to the actual demand that will be estimated, to the specific functional goals and the costs and the impact they will determine.

Eventually study compared on CBA basis the transport scenarios.

Primary goal is to provide massive public access to Westfield Milan Mall and adjoining HSR station, making it a real inter-modal hub of the Milan "East Gate", integrating it with the City and its Metropolitan Area and making the works a sustainable and upgrading development and environmental quality. This, in terms of means of transport, involves equally fast access to all mobility levels, from air traffic headed to nearby LIN airport but also to others rail linked or linkable, all rail services, local buses urban and extra-urban, long haul coaches, pedestrian and cycle paths and, secondary but however present, taxis, shuttle services and private car traffic to integrate via park&ride and kiss&ride parkings. That's the meaning of an intermodal hub.

Westfield Area location, along one of the main European railway lines, and not far from Milan's City Airport with its own metro station, favors the achievement of this primary target.

In order to make a real transport and mobility hub out of "East Gate", it is essential the construction of a new railway station, or the enlargement and upgrading of the present station, capable of accommodating all types of high-capacity rail services; the link between said station to the Linate airport terminal and to the new M4 metro station currently under construction, by extending the line or by means of some other type of link provided it is fast, frequent and attractive; a bus terminal for local and intercity buses and long-distance and tourist coaches; an efficient integration of the various transport infrastructures at a functional level, so that connections are fast, simple, intuitive, easy and comfortable in large common covered spaces fully available for use and architecturally appealing and attractive and the creation of a cycle paths network and pedestrian precincts which lead to and support the node, and which interact perfectly with the surrounding environment, also with the Idroscalo and the built-up area of Segrate.

MM set the described feasibility study. The purpose of this paper is to show how what it is supposed to be, and actually is the most engaging and expensive solution, M4 metro extension underground, can however be more advisable even in a relatively peripheral and not densely built contest such as area of interest.

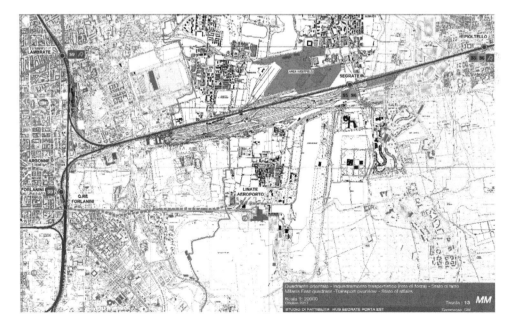

Figure 1. Westfield Milan – Porta Est Feasibility Study Contest (MM).

2 WESTFIELD MILAN

Described SF originates from large **Westfield Milan** mall, already under construction, in the Milan "ex-Dogana" (Customs) areas in Segrate. With more than 600,000 m² of total surface, over 420,000 m² covered and a shopping area of 235,000 m², Westfield Milan will be Europe's largest shopping center with over 300 sales points open for 363 days a year, more than 50 restaurants, a multiplex cinema with 16 auditoriums, a Luxury Village and car parks for 10,000 vehicles. Moreover, it will not only stand out for its size but also for its high quality at commercial level - with the presence of a distribution group such as *Galeries Lafayette* and of the leading fashion labels - and at architectonic level, which will make it an upgrade meeting and recreation center. As said, standing at some 2bn€ cost, *Westfield Milan* it's currently largest foreign investment in Italy.

Catchment area data are also impressive: A strategic location in the Milan metropolitan area, with a population of 5.3 million, at the gateway to Milan City, a barycenter of the Milan urban region (7,5 million inhabit. including metro areas of Varese, Como, Lecco, Bergamo and Brescia) and in the center of Lombardy (over 10 million inhabitants and between ⅕ and ¼ of national GDP). 6,4 million residents currently live within 90' from the center.

These numbers are the basis for Westfield Group's estimates of 25-30 million visitors per year and therefore in the order of magnitude of a permanent EXPO, an annual potential of 3 million tourists and 44,000 new jobs, without counting allied businesses.

In addition, with a turnover of € 5.6 billion in 2016, for luxury business Milan is ranked world's 8th, Europe's 3rd and by far Italy's 1st. In fact, Milan accounts for 2,25% of the world's luxury market (€ 249 billion) with just 0.07% of the population (including the metropolitan area).

This enormous potential justifies the investment, which accounts for a tax revenue of € 60 million as well as over € 270 million of VAT, while an increase of € 250 million in the annual earnings of the employed population is forecast.

Project has undoubtedly the critical mass to support and revitalize the strategic plan of the "Third Gate" of access to Milan, after the Trade Fair Center and the Rogoredo districts, linked to the development of the HS rail network. It must lastly be noted that the input plan and main concept of the works is Westfield's distinctive approach to the planning of its centers in the world, which is in the name of quality and sustainability rather than mere real estate and commercial speculation, as can be seen particularly in the "twin" Westfield Stratford in London, characterized by remarkable urban and environmental upgrading and the huge contribution of mass public transport, far more utilized than private traffic.

Figure 2. Westfield Milan Masterplan (Westfield Group).

3 PORTA EST HSR STATION

First strategical goal achieved by metropolitan planning has been to combine forthcoming Westfield Milan mall to a new HSR gate station, completing accessibility to Milan railway node working as a *hub* for all train services: high speed, long haul, Intercity, regional and suburban.

A real gate station is in fact currently missing on fundamental Milan – Venice railway axis.

Planned station will mostly consist in three levels, a ground track level, an underground underpass level and an elevated level with the real station atrium, working as an above mezzanine . Stations will have six tracks: two going through on fast line, two going through on slower historical line plus two terminal tracks for regional and suburban services having their terminus in place. 4 going through tracks will be served by three 410 m long side platforms while two terminal ones, slightly lower, will be served by a single central 250 m long platform.

It is important to notice that due to the multi-level approach, the station will have both underground and above ground pedestrian connections.

The new East Gate station is the hub on which the other transport elements are centered.

Hub is considerably close to Milan node and it is accompanied by one of the largest urban development plans in Europe, the bearer of huge numbers also in terms of visitors.

Station master plan will guarantee a close, easy and fast reciprocal access of the corresponding transport elements, namely East Gate railway station itself, M4 Linate connection, bus terminal and car parks.

Hub will be included in a tariff framework to guarantee the convenience of travelling through.

Station will interface with the urban context in which it is inserted, and especially with Westfield Milan, Segrate and last but not least the Idroscalo, also known as "Sea of Milan" and definitively one of its most remarkable parks.

As it's been planned by feasibility study but not projected in details yet, HSR station will so far fulfill following requirements:

- An adjacent Bus Terminal;
- A cycle-pedestrian connection which reaches the Idroscalo opening a second front;
- The use of part of the Westfield car parks as connection car parks for the node.
- An elevated above tracks plate with the function of a large, central distribution mezzanine suspended between two lateral station bodies;
- Two parallel platform underpasses at the two ends of the platform, for direct access to the car parks (west side) and to the Bus Terminal/former Segrate RFI stop (east side).

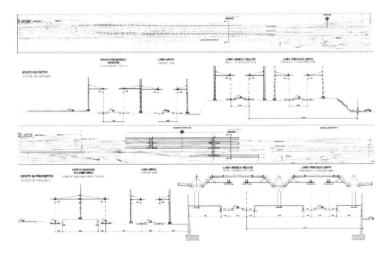

Figure 3. Porta Est Feasibility rail plan and cross sections (MM).

– The total, complete coverage and mechanization in both directions of all the connections between the railway station and the terminus of the connection system to Linate, the Westfield center and possibly the Bus Terminal and the Car Parks.

4   WHY UNDERGROUND?

Why did we have to move underground, choosing the metro system?

As a multi-scenario and multi-system feasibility study, Porta Est one has pointed out as said four typological alternatives to connect LIN/M4 to Westfield Milan/HSR station Porta Est in addition to M4 itself extension: Busway, Cable Car, Monorail.

Significantly, all of them are way less expensive than underground metro option.

Busway involves a total cost of some 78M€ considering a massive road empowering and enlargement (otherwise would be just 20-30M€), Cable Car just needs 51M€, Monorail (a light, slender one track shuttle system as monorails can easily get to same dimensional size than heavy metros) about 125 M€ while underground M4 extension peaks up to 220 M€.

Likewise, also construction time is way longer for a massive infrastructure such as an underground TBM excavated metro line: at least 5 years are needed while half of such a time is enough to complete all the other evaluated systems, above ground and largely prefabricated in the case of cable car and monorail.

An easy answer could be that metro system has a much higher capacity. With 600 pax/train (6 pax/m^2) and a 90" peak frequency M4 will in fact guarantee a maximum capacity of 24.000 pphpd. That's 12 times busway capacity, about 2.000 pphpd, 8 times 3.000 pphpd cable car one and 4 times monorail 6.000 pphpd one.

Quite a lot, sure. But the point is: did we *really* need all of this capacity?

Let's see. Demand analysis reported an estimated 21.000 – 26.400 passengers demand.

Per day. In both directions together.

Namely 2.100 – 2.700 pphpd, just 10% of M4 system capacity. Barely 15% considering a more comfortable 4pax/m^2 index, which reduces M4 capacity to 16.840 pphpd. It is true that all diametrical metro lines have a "Gauss-profiled" demand diagram, diminishing towards the terminals, however in this case two of three other system can perfectly fit such a demand. Theoretically, while practically all of them as a less performing system involves a lower demand estimate. So, basically, the answer is *no*, in absolute terms we didn't need all of this capacity.

Back to the main question, why moving underground by metro, are performances the key?

Not quite, in this case. Sure, M4 average speed is about 30 km/h on entire line while in the specific segment peaks up to 58 km/h while busway, cable car and monorail, commonly between 20 and 30 km/h on entire routes, are in this case limited to 30, 20 and 40 km/h respectively, much less than M4. However, as we're speaking about a 2,5 – 3 km route, also slower system is able to run it in a few minutes. Actually 372" (little more than 6') the busway, 440" (7'⅓) cable car and 245" (basically 4') monorail, while M4 just makes the trip in 160", 2'⅔. Way quicker, but on such a small scale means few minutes, a bunch of seconds less, definitively not decisive.

Besides, the question is not that much why choosing the metro, the question is why moving underground firstly. If total capacity and speed were crucial issues, what would prevent to build a M4 extension above the ground, at grade or in viaduct like other evaluated systems?

Answer at that point could be that a metro system is less flexible than a busway or a monorail, therefore above the ground in urban areas can be problematic. But apparently not in this case.

In despite of being the contest located in Milan's metro area eastern first ring, its urban density is way below the average of surroundings. Milan is a very dense city, standing its central Municipality at 7.700 inhabit./km^2. Also metro area as a whole has an average urban density close to 2.700 inhabit./km^2. Local eastern first ring Municipality of Segrate has too an over than 2.000 inhabit./km^2 density (2.057). But presence of LIN city airport, Idroscalo and Forlanini Parks plus *Smistamento* (literally "shunting") railyard has determined the

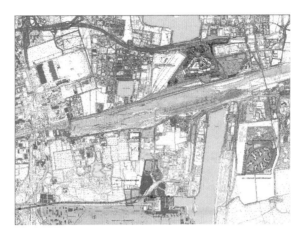

Figure 4.   Westfield Milan – LIN/M4 area (MM).

persistence of considerably wide empty spaces around, apparently perfect to be exploited to work above ground, and save money.

But empty doesn't mean free. At a first glance, all empty urban spaces can look suitable for massive infrastructural works difficult to place in urbanized areas and expensive to place underground, but often a more detailed study of the contest, main base of every infrastructural feasibility study, helps to figure out several important restrictions.

In the specific case, airport proximity sets mandatory height limits all around. This is a strong and basically all over significant limit to all viaduct works, that can't involve too tall structures. There are and there can be viaducts, but overpassing at grade infrastructures not far from minimum allowed standards. Limits are even stricter when energy supply is above, even if his is not the case of 3rd rail supplied M4.

That's not all. Area is split in north and south part as crossed by Milan-Venice rail line, locally deploying an up to 500 m width due to adjoining yard *Smistamento*. A some km long and some hundreds meters wide obstacle impossible to cross at grade. And overpassing it means having at least 7/8 m differences between rail levels.

Enough to stay within air limits. But what's not clear on a simple survey base is the fact a wide part of this yard will evolve into a *TERALP* freight terminal in the next years. And when you have a freight terminal you have overhead traveling cranes. Overpassing them by viaduct is a though job. Impossible with local air limits.

Moreover, a viaduct or even at grade infrastructure running north-south west off Idroscalo would cut the green continuity between Forlanini Park and Idroscalo Park, soon to be re-enhanced by LIN masterplan, which will ease pedestrian permeability to the "Sea of Milan". Of course it's possible to avoid it, moving the path far from it and the airport, gaining precious additional meters to use in height. But for this you need a more flexible and light infrastructure.

LIN/M4 area is close to some military areas, very difficult to interfere in, as it's been proved during the still going on construction of local M4 terminus. Adding new infrastructure totally above ground would had made things much more complicated, especially in cable car option. Monorail terminus on the other hand, placed as a "bridge structure" between two parking lots, would limit height of under passing vehicles, which can be problematic in the case of fire brigade emergency trucks that must be able to go everywhere and quickly in an airport area. And on the opposite terminus, Westfield Milan, economic booster of the whole plan, the interference with the area would be maximized as viaduct based depot can't be placed anywhere else. Not a welcomed fact for investors.

Back to rail yard, the some 500 m wide overpassing problem is common to all systems if you decide to stay overground. For buses too. That's why it would not just be a busway. It'd be a freeway viaduct, much more expensive than a simple busway at grade. And, even more important, viaduct would just be a part of a totally new and enlarged highway opposed by local authorities

Cable car was suitable according to the study. But, in spite of its lightness, it's rigid in terms of route. Which meant two things: moving terminus far from connections at LIN/M4 and Westfield/HSR station, worsening correspondences, and put a severe limit to infrastructure's capacity. 3.000 pphpd and not more. And slowly. Forever. Plus many people can have psychological problems to use this mean of transport to commute to/from HSR station, metro or airport. What in the case of storms or simply wind? Never a big problem in Padana area, and cable cars are among world's safest infrastructures, but for many people can be very scary. A lot more and more frequently than in the other more common means, that to people look basically like trains or buses.

Speaking about passengers issue, it is true that in terms of absolute capacity, basically all mean of transport were able, or almost able, to carry M4 expected loads. And all of them with no exception were able to carry the expected load for themselves as systems. But that's the point: not all systems have the same appeal. Metro is more attractive and especially in this case as the connection LIN/M4 – Porta Est HSR/Westfield by metro means NO intermediate reloading. No additional correspondence, always discouraging public as people are already headed to a correspondence (rail-metro or metro-air).

So, according to demand esteems, if M4 would catch some 24.000 passengers per day, monorail would just got 10.000, cable car 8.000 and busway 7.000, that's a real difference, as we're not simply talking about suitability of new infrastructure, but about its success. This is a plan involving huge investments, therefore revenue is even more important.

Eventually, combination of all these problems makes evident how the "empty" area is definitively not easy to exploit. Even considering a M4 metro option above ground, that would just increase other systems problems due to larger size and minor flexibility, compared e.g. to monorail. There wouldn't be a depot – already foreseen close to opposite M4 west terminus *S. Cristoforo* – but impacts would increase, especially close to Linate station and nearby military area (avoided by monorail) without having therefore a dramatic drop of costs.

On the other hand, moving underground other systems is simply no feasible: total waste of money for busway and monorail, having basically same civil works costs for way lower capacity and performances and yet maintaining intermediate transfer, completely absurd for cable car, even turning it into a funicular. Summarizing:

- Busway wouldn't be that cheap as a 600 m massive freeway viaduct should be built over a trainyard and expected demand would be the lowest, plus the reloading break;
- Cable car would be even more limited and far from fundamental correspondences, plus presenting building site and psychological problems whose solution can't be given as granted and additional transfer needed too;
- Monorail would maximize impact on Westfield Milan site and create problems to fire brigade vehicles in airport area, plus it's however expensive: more than a half of metro but having ¼ of the capacity and ⅓ of expected demand, plus again the intermediate passengers reloading.

A popular Italian proverb quotes "*Chi più spende meno spende, e chi meno spende spende più*" (Who spends more spends less, and who spends less spends more) to remind that sometimes choosing the cheapest solution can be cheaper at the moment but could be more expensive in the long run. This is probably a case.

Metro network should reach all HSR gate stations without additional transfer and no severe capacity limits. M4 is the only evaluated option able to assure these two important strategical goals. Worth spending 2 to 4 times more, having however higher revenues, directly (M4 ticketing) and indirectly (Westfield Milan, HSR station and LIN spin-off).

Moreover, going underground is the only choice which makes possible to give as granted functional and technical feasibility, impossible to assure in all other options. Porta Est/Westfield Milan case is a typical example of how a well-tested engineering combination like TBM excavated M4 metro line currently under construction can solve complex problems assuring both achievement of highest performances and importance of current real estate developments.

Figure 5. Height restrictions due to air traffic (left) and environmental restrictions (right) (MM).

5 M4 METRO LINE

Currently under construction, M4 line will be fully operational at the end of 2022, with Forlanini-Linate segment opening in 2021.

M4 line is a light, entirely automated (driverless) underground metro rail line with 50 m length standard for trains and platforms.

It runs east-west through the center of Milan from S. Cristoforo FS to Linate Airport, it is 15,2 km long with 21 stations and a line depot of 140.000 m² able to host the 47 trains fleet foreseen, which consist in the number needed to guarantee the planned peak frequency of 90".

The trains, convoys of 4 intercommunication carriages without a driving cabin, have a maximum capacity of 600 pax (6 pax/m²), consequently the system has a theoretical maximum capacity of 24.000 pphpd, with a maximum on-line speed of 80 km/h and a commercial speed of 30 km/h.

Due to such an high performance profile, M4 line will make possible to cross Milan City east-west in just 29' and connect LIN city airport to historical center in some 10'.

Cost is pretty high, about 2bn€, namely about 132M€/km, including depot, fleet and planning.

The extension of the M4 line represents most obvious and immediate manner, from a theoretical viewpoint, to resolve the Linate M4 - Segrate East Gate Westfield connection. As of 2022-2023 the M4 line will link Linate to the city center, greatly increasing its accessibility and

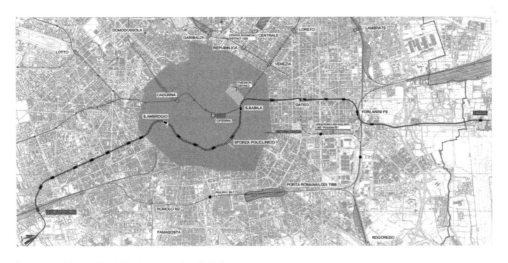

Figure 6. Metro line M4 chorography (MM).

urban "proximity", while even earlier, as of 2020, the Westfield shopping center will rise, the largest shopping center in Europe, and these two important centers will be less than 2 km apart as the crow flies, separated by the green area of the Idroscalo and by the railway switching yard. Obviously, it is also the most expensive and complex option.

Extension is 2,6 km with one new station, new Porta Est terminus, identical to the current terminus being completed at Linate: there would be a considerable distance between stations but still less than the distance between Linate Airport and the Forlanini district.

A double-track would be built in a single TBM excavated tunnel lowered near the new Porta Est station when the work-site is opened and recovered at the end of the current shunting siding.

The route starts from the end of the current shunting siding, which, as for the LIN station, has a track about 10 m below the level of the surrounding land. From here the track continues in a straight line along a special chamber created by an open-cast excavation, for about 200 - 250 m, within which the track level falls to a sufficient depth to allow the use of the TBM, arriving from the opposite side. The chamber implies open-cast working, therefore with an impact on the surface, along the north bank of the "Linate Branch" of the Idroscalo, stopping at the terminal. Working in the chamber, and the recovery of TBM's milling head, could be contextualized with new "Piazza Idroscalo" construction envisaged by SEA Masterplan for LIN. At the end of the chamber, the track turns to the north with a radius of 200-220 meters into the tunnel excavated by the TBM, then continues in a straight line for about 1,5 km at a depth of 15 m, then with a similar curve it turns eastward passing under the FS switching yard, to then reach the Porta Est station which is positioned parallel to the railway line and which will have to have be at a depth as close as possible to the surface in order to optimize the connections with Westfield, with the railway station platforms and bus lines. The "infrastructure package" of terminus station + shunting siding is completely similar, as regards size and electro-railway equipment, to that under construction at Linate, with a 238-metre-long shunting siding. Both station and shunting siding will be created with a cut&cover excavation.

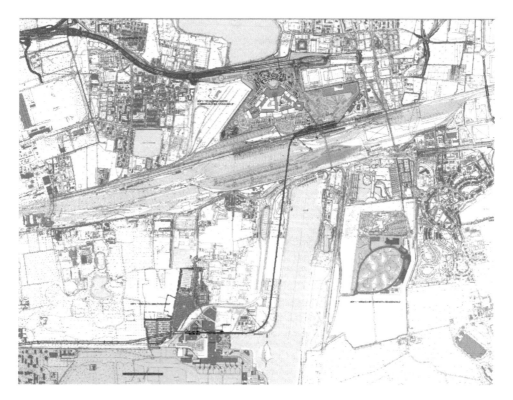

Figure 7. M4 underground extension from LIN to Westfield Milan/HSR Porta Est railway station (MM).

The experience of the line currently under construction from Linate to San Cristoforo offers valuable ideas to contain, within the limits of the values of a large work, construction costs and the engineering commitment. It is here necessary to keep the line as near the surface as possible, although within the limits allowed by the TBM, improving the accessibility and the economy of the stations, thanks to the volumes reduction.

With regard to safety facilities, safety criteria have been adopted for the line currently under construction based on the prescriptions issued during the approval procedure, while the more recent regulations on fire prevention (Min. Decree 21/10/2015 "Approval of the technical fire prevention regulations for the design, construction and operation of underground railways") could allow for savings by reviewing the number of line facilities.

With a 171" only additional time from LIN to Porta Est, City Airport will be reachable in just 3' and city center in less than 15'. M4 extension requires 4 more trains, with 1 additional emergency train, bringing the total of the line to 51.

Pros and cons of the M4A system can be summed up as follows:

– Very fast and very frequent service;
– Very high system capacity;
– No intermediate transfer between Linate and Porta Est;
– Direct and streamlined line;
– Higher costs and greater complexity;
– Longer construction time.

A definitively positive balance, which makes this solution not only suitable, but perfectly matching the large scale strategic targets to be achieved.

6   CONCLUSIONS

Metro excavating experiences have taught that what sometimes seems to be a saving money issue at first sometimes comes not to be likely. A typical example, especially in Milan's and other Italian cases is the surface/underground choice. In contest such as this paper's object it's easy to be fascinated by the possibility of bringing the metro on surface, once finally downtown high density is over and open spaces are available between park and airport. Perfect occasion to exploit the 1 to 10 cost ratio in favor of surface. But such a ration ends up being only theoretical. In the reality, many times several restrictions and obligations occur, quickly consuming even such a huge cost advantage making in addition working time longer and less secure, due to the several interferences.

It is not always a smart thing: it has to be admitted Italian legal framework regarding infrastructures is way too much muddled and unimaginatively restrictive, so much to make smart choices not convenient. This applies very frequently to at grade solutions for rail infrastructures, making the path – and especially the stations – unreasonably more massive, overloaded and redundant than anywhere else in Europe and frankly without a real technical reason. Recent TBM tunnelling development partially deepens its roots in this issue too.

However as in this case the metro option was motivated not only by demand needs but also by a matter of image, the underground TBM choice helped in bypassing major obstacles which would have reduced surface segment to a not significant one in terms of costs saving but wide enough to stop park continuity.

REFERENCES

MMspa (Andrea Bruschi, Sandro Capra), October 2017. Segrate metropolitan hub – Milan East Gate, technical and economic feasibility project.
Westfield spa, June 2015. Studio di Impatto ambientale (Environmental Impact Study).
SEA, November 2016. Linate – masterplan aeroportuale (LIN Airport Masterplan)

Tunnels and Underground Cities: Engineering and Innovation meet Archaeology, Architecture and Art, Volume 10: Strategic use of underground space for resilient cities – Peila, Viggiani & Celestino (Eds)
© 2020 Taylor & Francis Group, London, ISBN 978-0-367-46878-1

Shafts and drainage tunnels as a sustainable and preventive measure to mass movements of saturated soils in Bucaramanga, Colombia

P.F. Castillo & G. Pardo
Colombian Tunnelling and Underground Works Association - ACTOS, Bogotá, Colombia

ABSTRACT: The city of Bucaramanga is located in the northeastern zone of Colombia. It has 1.1 million inhabitants and a growing economy that has generated the migration of approximately 80,000 people in extreme poverty, with a non-urban planning in its north zone. This area presents a potential risk of mass movements due to saturation of residual and conglomerate soils and to high seismic activity. This article presents the design of three drainage tunnels of 1.1 km with a cross-section of 7 m² each and twenty shafts for groundwater collection from 26 m to 63 m in depth, as the optimal evaluated solution in terms of the greatest economy and sustainability and the lowest socio-environmental and constructive impact. Having as fundamentals challenges the combination of the physical space, the tunnel's portals and the presence of a nearby river in a dense urban area, finite element programs were used for this purpose.

1 INTRODUCTION

The city of Bucaramanga is the capital of the department of Santander, Colombia (Figure 1), it is located in northeast of the country on the Eastern Andes Mountain Range, and it has 1.1 million inhabitants. It has a diversified and thriving economy, which places it as the city with the lowest level of inequality in Colombia (Gini coefficient = 0.43). This, along with the forced displacement caused by the Colombian armed conflict, has stimulated during the last decades the migration of approximately 80,000 people in poverty who seek to improve their living conditions in this prosperous city. This vulnerable population has been settling illegally, disorderly, without urban planning or access to basic public services in vacant lots in the Northern Zone of the city.

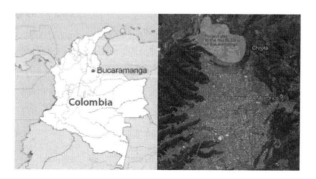

Figure 1. Project Site located in the Northern Zone of Bucaramanga, Colombia.

The geological conditions in this zone correspond to a colluvial deposit that presents historical problems of mass movements that affect the population, being the areas of steep slopes the most prone to this problem. The saturation of the residual and conglomerate soils generated by the intense rainy seasons, the high seismicity, a product of the seismic nest of Bucaramanga, where South America and the Caribbean tectonic Nazca Plate collide and the fault system adjacent to the zone of the morphological modifications generated by anthropic changes of cuts and fillings increase the level of occurrence of these phenomena. For this project, an area of study with 1km² an altitude between 640 meters above sea level and 880 meters above sea level has been established. This area includes 2412 properties with approximately 12,060 inhabitants.

2 DESIGN ASPECTS

The threat of mass movements was analyzed with the Rocscience Slide 7.0 Software for different scenarios with critical factors being earthquake and rain. To determine the earthquake that could generate a movement in total mass, the probability of total failure before a seismic event was estimated with a specific gravity acceleration value (based on fault length, epicentral distance, fault systems, recurrence curves, return period and probability of occurrence), for a useful life period of 50 years through the total probability theorem. The parameters were obtained from the historical record of seismological stations (1952–2017) of the Colombian Geological Service -SGC and a particular seismic study in the area (based on the Cornell Model). As a result, a critical earthquake was obtained for the Suratá fault with a 6.7 MW magnitude and acceleration in the study area of 330.88 Gal (for Pseudostatic acceleration coefficient $K_{st} = 0.26$, Return period T = 475 years and Acceleration in rock $a_t = 0,26$ g). With regard to rain as a critical factor, a hydrological and hydro-geological characterization of the study area was carried out with the historical record (1983–2003) from the Institute of Hydrology, Meteorology and Environmental Studies - IDEAM and the increase of the water table was considered for a maximum rain occurred during the last 20 years. The results indicated that, only in scenarios with a seismic and rainy event, could there be a large mass movement throughout the study area, so priority is given to proposing a solution for local mass movements. These were determined with inclinometry measurements, showing gradual movements in the shallow part that vary at a rate of 1.6 cm/year to 2.5 cm/year (Figure 2).

In general, the mass movements occur due to an increase in the pore pressure of the soils and the decrease in the shear resistance of the saturated materials generated by the rise of the water table (which varies in depth between 1 m and 5 m), associated with long periods of rain. In order to solve this problem, pre-feasibility designs were proposed for two solution alternatives: 1. Penetration underdrains and 2. Shafts and drainage tunnels. The alternatives are mainly based on the following components: hydro-geology, hydrology, geomorphology, geology, geotechnics, project management and socio-environmental analysis, among others. A useful life of more than 50 years was defined as a fundamental design parameter, in order to guarantee its sustainability over time.

The solution alternatives were evaluated in a matrix, identifying the impacts generated by the construction and execution activities of each alternative, on the different social and environmental elements. The impacts were rated and ranked in order to establish the elements,

Figure 2. Housing affected by mass movements.

components and systems that would be most affected in the construction. The results indicate that the optimal alternative in economic, operation and maintenance terms, with minimal constructive and socio-environmental impact corresponds to the shafts and drainage tunnels. This alternative has a higher initial cost, but over time, it is forecasted with a minimum investment in operation and maintenance of the structures, unlike penetration underdrains which are exterior works exposed to effects by critical factors (rain and earthquake) and with need of periodic maintenance due to constant surface movement. According to the above, the given alternative provides the stability of the colluvial deposit in the medium and long term, by lowering the groundwater levels, it presents minimal structural affectations before a seismic event and, in its conception, it has special respect and commitment with society and the environment.

For the chosen alternative and based on positive and negative weighting, a socio-environmental management plan was developed considering the recommendations of the National Environmental Licensing Authority- ANLA. In terms of socio-environmental management, were proposed programs such as: socialization of the project, community participation, hiring of local labor, training and education, solid waste management, wildlife management, saving and efficient use of water and energy. A cash flow analysis was carried out, and in order to organize the official budget of the project, a Work Breakdown Structure (WBS) was developed.

3   METHODS AND PROCEDURES: SHAFTS AND DRAINAGE TUNNELS

The plant layout determined for the shafts and drainage tunnels system is subject to their length, depth, location and incidence on the slopes where the tunnels portals works are located. With the information of land use, topography, geology and geotechnics, the different possibilities of access to the area of the works were analyzed, so that there was an optimal layout of the spaces intended to house the control facilities and a feasible connection with the access path of the tunnels. Likewise, an area where operations in excavation processes are minimized avoiding areas identified as potentially unstable or with existing urban interferences, and which allow the orientation of the underground sections in such a way that they cross as perpendicularly as possible the stratification or foliation planes, the effort zones and the geological faults previously anticipated. Based on the above, two alternatives of plant layout and profile were proposed to satisfy all the mentioned points.

With the hydrogeological modeling in Modflow, it was possible to comprehensively understand the underground flow system in the study area, which is underlying to a colluvium with a substantial extension and the flow system is mainly fed by the waters coming from the plateau and is drained to the North in the Suratá river. The model allowed to quantify the expected flow of groundwater collection in the permanent state and the efficiency with respect to the draw-down of the water table of both alternatives. Three scenarios were evaluated in the model and the results show that through Alternative 1 (Figure 4) the highest flows are captured. This alternative consists of 20 shafts that vary from 26 m to 63 m in depth and three drainage tunnels: DT 1.1, DT 1.2 and DT 1.3 with lengths of 367 m, 422 m and 335 m respectively, and estimated infiltration rates of: 6.0 L/s, 13.3 L/s and 10.8 L/s respectively.

To define the dimensions of the access portals to the DT1.1 and DT 1.2 tunnels that are close to the Suratá river bed and its tributaries, its flow rates and maximum levels were analyzed and a maximum flood level of 740 meters above sea level was defined, which allows to cross the stratum to be drained. With regard to DT 1.3, its access portal is located at 762 meters above sea level, the most suitable to cross the stratum to be drained.

To determine the vertical alignment of the tunnels, the first design criterion is a minimum slope that guarantees capturing the largest groundwater, being defined between 1% to 2%. Regarding the geometry of the shafts, its depth is provided up to the intersection with the corresponding tunnel crown and its diameter $\varnothing = 1.8m$ is the minimum necessary to proceed with a construction using the manual method of excavation. Its location in the plant was defined so that it did not interfere with urban infrastructure. The geometrical design of the service cross section of the tunnels covers aspects such as: constructive facilities that guarantee adequate construction times and excavation conditions, removal of material, ventilation, energy and operation of the equipment, safe for support placement. Additionally, the necessary width for the provision of maintenance equipment, flow meters and any other type of technology that is required to install to take control of the water collection process. With that in mind, a horse-shoe section of 7 m²is foreseen (Figure 3).

Regarding the groundwater collection system, it has been designed by drain holes from the tunnels and shafts. For the tunnels, radial rings of drain holes of Ø1" in TZ 9 to 3, with a length of 3m and spaced every 3 m, were defined. For the shafts, six radial drain holes of Ø1" with length of 1 m and spaced every meter were defined, for each shaft. To avoid fine material migration, all drain holes were designed with slotted plastic pipe and non-woven geotextile.

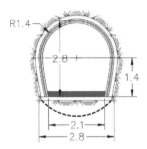

Figure 3. Typical section of drainage tunnels.

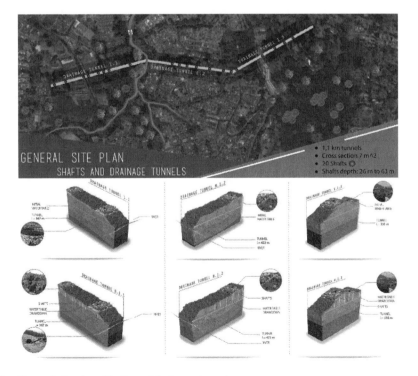

Figure 4. General site plan: Shafts and drainage tunnels.

3.1 Geology and Geotechnics for shafts and drainage tunnels

From the analysis of the predominant geoforms in the Northern Scarp of Bucaramanga, as well as from the information recorded in previous studies, technical visits and the analysis of multitemporal information, it was identified that the current movement zone is located on a large colluvium, the product of one or several ancient mass movements which were deposited on a stepped slope. Once the general layout of the local mountain relief was defined, the information of the stratigraphic columns, trenches, test pits and geophysics was used to define the geometry and thickness of the materials, in particular the colluvial component on which cuts and fillings were made to construct the existing homes in the study area. As a result, stratigraphic profiles that group the materials into homogeneous zones from the point of view of expected characteristics and behavior were generated. In this sense, for the general profiles, the colluvium (Qd) was modeled as a highly heterogeneous unit, the Upper Organ Member (QbO2), the Lower Organ Member (QbO1), which was considered as a single segment deposited on the rocks of Tiburon (Tt), Bocas (Jb1) and Diamante (Cpd) formations (Table 1):

Table 1. Formations of the general stratigraphic profiles.

	Name of the formation	Description
	Diamante Formation (Cpd)	Lower Cpd: Muddy sandstones that vary in size from fine grains to reddish gray grains with interlayered siltstones. Intermediate Cpd: Dark gray lodolite and limestone insertions of the same color. Top Cpd: Dark gray, slightly clayey limestone with thin claystones insertions and gray to grayish-red sandstones
	Tiburon Formation (Tt)	Calcareous, lithic, clasto-supported conglomerates, with fragments of cobbles and pebbles (2-10 cm) of limestones predominantly, with fragments of sandstones, cherts and volcanic rocks, within a sandy-calcareous matrix, gray and fine-grained.
	Bocas Formation (Jb1)	Insertions of siltstones, sandstones and calcareous claystones, greenish gray and dark gray, greenish gray siltstones, with calcareous nodules, greenish gray conglomerates and dark gray claystones, fossiliferous, greenish gray to reddish gray, slightly calcareous siltstones. Towards the upper part, there are thin layers of volcanic rocks.
	Lower Organ Member of the Bucaramanga Formation (QbO1)	Predominant gravures clasto-supported to matrix-supported with insertions of slightly gravel-like sands. They come mostly from metamorphic and igneous rocks
	Upper Organ Member of the Bucaramanga Formation (QbO2)	It is similar to QbO1 in the relationship between sandy gravels and gravel-like sands, the gravel fraction generally prevailing. The main difference is the change in the composition of the clasts, evident in the overwhelming presence of clasts of sedimentary origin and in a smaller proportion there are some clasts of igneous and metamorphic rocks.
	Organs removed from the Bucaramanga formation (Qd)	It is established as a zone of colluvial deposits.

With regard to structural geology, through preliminary photogeological interpretation, structural and stratification data taken during all field trips, surveys, trenches and test pits and the results of seismic tests, the presence of two fault systems were established within the study area: 1. Bucaramanga Faults System (B2, B3 and B4) and 2. System of Transverse Faults (TN2, TN2, TN2-3 and TN3). The layout of these fault systems in the longitudinal profiles of the drainage shafts and tunnels was taken into account, since they explain the relative displacements of contact between formations. In Figure 5, Figure 6, Figure 7 and Figure 8 the longitudinal profiles of the drainage tunnels are shown, with slopes ranging from 1% to 2%. A flood protection zone is available in DT 1.1 and DT1.2 of 8 m and 7 m respectively; after this, there is an access road and a maneuvering yard. It was decided to implement a false tunnel that begins past the access road and the maneuvering yard, to minimize interventions in excavation processes. For DT 1.3 no flood protection zone was established and consists only of an access road, a maneuvering yard and a false tunnel. All the necessary cuts were made to analyze the maximum coverage measured from the key of the tunnels that allow to analyze the efforts, deformations and subsidence through finite elements programs.

The definitive design of the tunnel portal zones covers the minor disturbance to the land, creating a false portal and avoiding the big earth movements in each case, situation which additionally minimizes the effects on the environment.

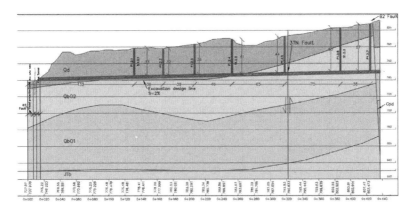

Figure 5.   DT 1.1. Longitudinal profile.

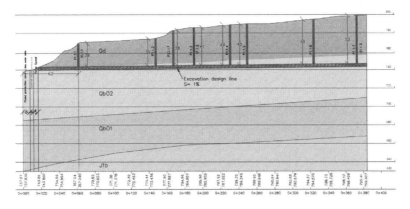

Figure 6.   DT 1.2. Longitudinal profile.

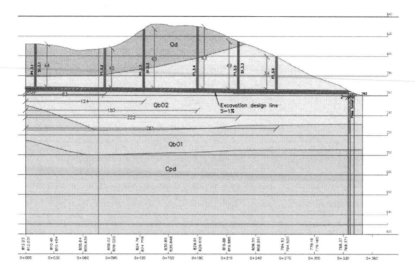

Figure 7.   DT 1.3. Longitudinal profile.

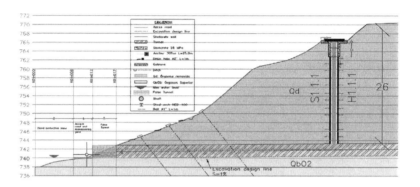

Figure 8.   D.T. 1.1 Inlet portal profile (Typical scheme).

3.2   Numerical modeling with finite elements method for the verification of the shafts and drainage tunnels support

For the modeling, the information previously obtained from the geotechnical and geomechanical sectorization of the land was used, the support sections for tunnels and shafts were defined and finally it was verified if the reinforcement is adequate and sufficient to prevent collapse or occurrence of subsidence through the Rocscience Phase 2 V. 8.0 Software. Three types of terrain were planned for the tunnels and shafts (Table 2) In the same way, 3 support sections associated with each type of terrain were designed for tunnels, and for the shafts a standard support was designed to guarantee stability throughout its depth.

The modeling carried out show satisfactory results (Figure 9) with total settlements around 1.2 cm to 4 cm and resistance factors of the shotcrete and bolts around 1.26 to 6, a value that guarantees that the support defined above is adequate and sufficient for the system and prevents unacceptable deformations in the determined useful life.

The subsidence was analyzed in relation to the construction of the shafts and tunnels system and the one generated in the medium and long term with the draw-down of the water table. The results of the models indicate that there will be no significant settlements that place the surrounding infrastructure at risk. The draw-down of the water

Table 2. Types of terrain.

Description	Total Length (m)	Type of Terrain	Length (m)	Support sequence
DT 1.1	367	Type V Land	274	• Shotcrete thickness = 0,10 m (f'c = 28 MPa) • Spilling pins Ø1" if they are required with a length = 2 m, they will have a radial spacing of e = 0.35 m, distributed at 45 degrees on both sides of the vertical axis, i.e. between TZ $10^{1/2}$ to $1^{1/2}$ • Type A Ø1" Bolt in the entire section with length = 2 m with a radial spacing of e = 0.5 m. • HEB-100 steel arch or equivalent, spaced = 1 m. • Shotcrete e = 0.05 m (f'c = 28 MPa)
		Land Type Soil	93	• Shotcrete thickness = 0,10 m (f'c = 28 MPa) • Type C Ø1" bolt distributed at 45 degrees on both sides of the vertical axis, i.e. between TZ $10^{1/2}$ to $1^{1/2}$ of length = 6 m with a radial spacing of e = 0.25 m. • HEB-100 steel arch or equivalent, spaced = 0,75 m. • Shotcrete thickness = 0,05 m (f'c = 28 MPa)
DT 1.2	422	Type V Land	123	
		Land Type Soil	260	
DT 1.3	335	Fault Type Land	39	• Shotcrete thickness = 0,10 m (f'c = 28 MPa) • Type C bolt Ø1" distributed between TZ 9 to 3 of length = 6 m, overlap of 2 m and with a radial spacing of e = 0.3 m. • HEB-100 steel arch or equivalent, spaced = 0,50m. • Shotcrete thickness = 0,05 m (f'c = 28 MPa)
		Type V Land	303	
Shafts	884	Land Type Soil	32	
		Type V Land	289	• Shotcrete thickness = 0,10 m (f'c = 28 MPa)
		Land Type Soil	**595**	• HEB-100 Arch or equivalent, spaced = 1 m • Type A bolt Ø1 "L = 1 m • Shotcrete thickness = 0,10 m (f'c = 28 MPa)

table due to the construction of the project is minimal, given that the rate of descent is low and will be controlled by the constant infiltration flows. A constructive procedure was formulated to reduce the possibility of causing damage to public services buildings or infrastructure. That is why excavation advances of maximum 1 m were established, duly supported with shotcrete, rock bolts and anchors. This procedure adequately controls the displacement in the front of the tunnel, the convergence of the side walls and the possible displacements of the support due to deformations in the lower areas of the tunnel.

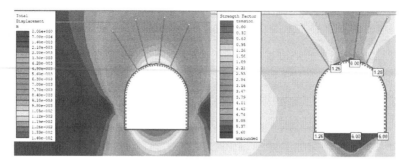

Figure 9. Total displacements and resistance factor of the model.

3.3 Construction system

The construction system defined for the shafts and drainage tunnels system is a combination of the manual system of excavation supported by light machines since it will allow to continuously evaluate the conditions of stresses and deformations, adjusting if necessary, the specific support to the advance. In the upper part of the same, a concrete ledge was built that was used to install the equipment and as surface protection; for operation, a reinforced concrete cover was placed on the ledge.

4 RESULTS AND CONCLUSIONS

This article highlights the importance of the use of underground space for the solution of social problems in a country with limited resources and affected by internal conflicts; it also prioritizes the vulnerability of the population and the importance of immediate attention by government bodies, engineers and city planners; through the implementation of a prevention project in the face of the threat of mass movements with a useful life of more than 50 years, socially and environmentally sustainable with a low impact on the daily life of the population.

This project is part of the Comprehensive Plan for the development of the Northern Zone of Bucaramanga. The total cost is USD $5,907,403 and its construction will take 18 months. Programmatically, it is the first project to be implemented, as it guarantees stability in the face of mass movements in the study area and ensures the sustainability and durability of the other complementary projects focused on improving the quality of life of the community; among them we have: total coverage of public services for the population, improvement, rehabilitation and construction of access roads, investments on educational and health infrastructure, adaptation of green and recreational areas and architectural renovation, etc.

One of the fundamental benefits of the project is that the initial objectives are met, taking advantage of the gravity force, water pressure and underground space. There is no power consumption for the processes since the drainages are natural, obtaining over time a draw-down in the water table that guarantees a decrease in pore pressure, an increase in the angle of friction of the soil and therefore, a tendency to minimize local mass movements by 85% with respect to the rata measured.

Due to the nature of this article, the results submitted here are the most representative in relation to the shafts and drainage tunnels design, although there are more results related to the other mentioned areas and analyzed throughout the document.

The new social approach of Underground Engineering will allow Colombia to solve many of its problems associated with the difficulties imposed by the Andes Mountain Range. It will improve the mobility, the hydraulic use, the control of landslides, among other aspects. It will demand a great technical effort from the new generation of tunnellers in order to deepen knowledge of the Andean geology and geotechnics aspects.

ACKNOWLEDGEMENTS

The authors are grateful to Universidad Industrial de Santander – UIS Technical Equipment, Bucaramanga's Metropolitan Area and Bucaramanga Mayor's Office for their permanent support and supply of all the needed information in this paper. Finally, the support from the Colombian Tunnelling and Underground Works Association – ACTOS.

REFERENCES

Suárez, J. 1998. Deslizamientos y estabilidad de taludes en zonas tropicales. Bucaramanga: Ediciones UIS.

Suárez, J. & Porras, H. 2017. Estudio de amenaza, vulnerabilidad y riesgo por movimientos en masa del Sector Norte de Bucaramanga. Bucaramanga: Universidad Industrial de Santander – UIS, Grupo de Investigación Geomática, Gestión y Optimización de Sistemas.

Pardo, G. 2001. Construcción del Túnel alterno de Usaquén, Falla de Usaquén: Aspectos Técnicos y Constructivos. Bogotá D.C. Sociedad Colombiana de Ingenieros.

National Environmental Licensing Authority- ANLA 2015. Resolución 0114 de 2015, se adoptan los términos de referencia para estudios de impacto ambiental.

Colombian Geological Service –SGC 2017. Historical record of seismological stations. Bogotá D.C.

Colombian Geological Service –SGC 2015. Guía Metodológica para Estudios de Amenaza, Vulnerabilidad y Riesgo por Movimientos en Masa. Escala detallada. Bogotá, D.C.

The Institute of Hydrology, Meteorology and Environ-mental Studies – IDEAM 2003. Historical record (1983–2003).

Tamez, E., Rangel, J.L. & Holguín, E. 1997. Diseño geotécnico de Túneles. México D.F: TGC Geotécnica S.A. de C.V.

Novak, P., Moffat, A.I.B., & Nalluri, C. 1994. Hydraulic structures. Nueva York: Mc Graw Hill.

Dolz, J., Gómez, M., & Martin, J.P. 1992. Inundaciones y redes de drenaje urbano. Barcelona: Colegio de Ingenieros de Caminos, Canales y Puertos de Madrid.

Goodman, R.E. 1980. Introduction to rock mechanics. New York: John Wiley & Sons.

Wood, D.F. & Morgan, D.R. 1993. Shotcrete for Underground Support IV. New York: American Society of Civil Engineers – ASCE.

López, C. 2011. Manual de Túneles y Obras Subterráneas. Madrid: Universidad Politécnica de Madrid.

Memorias, XI Seminario Andino de Túneles y Obras Subterráneas – ACTOS 2016. Bogotá D.C.

Tunnels and Underground Cities: Engineering and Innovation meet Archaeology, Architecture and Art, Volume 10: Strategic use of underground space for resilient cities – Peila, Viggiani & Celestino (Eds)
© 2020 Taylor & Francis Group, London, ISBN 978-0-367-46878-1

Effective use of underground space for sustainable cities

S.W. Chiu, K.H. Lai, K.W. Cho & S.P. Lai
WSP (Asia) Limited, Hong Kong

ABSTRACT: The land resource in a well-developed city is limited. One possible to create more space is moving development at grade into underground space. For example, metro systems, underpasses and road tunnels are some common practice to achieve effective transportation without occupying the space at grade. Also, utility services are laid underground individually in some cities. Combined services tunnels can be considered to organize the underground utility services better. In addition, by putting sewage treatment and refuse transfer facilities underground, it can provide a better environment to neighboring areas. With reference to projects in China and Singapore, this paper will discuss the recent trend on the effective use of underground space and highlight the major design considerations to achieve a good result. An integrated approach and proper planning of the construction work to minimize the impact due to the construction of underground facilities will be addressed.

1 INTRODUCTION

Rapid population growth has led to increasing human-land conflicts because of the growing demand for resources and infrastructure. Developing underground space is one of the measure to relieve these problems and improve the sustainability of cities. Underground space can be utilized in various aspects, such as transportation and public utilities.

For transportations, metro systems and road tunnels are examples that can be constructed underground without occupying limited land resource. In order to be more effective, the transportation network can be integrated with the developments nearby. Public utilities that require large area, for example, water treatment facilities, can be constructed underground such that more space at grade could be released. Their effects to the surrounding environments could be minimized simultaneously.

On the other hand, road excavation and other issues might arise due to maintenance of utility services. To solve the problem, combined service tunnels, which are man-accessible tunnels incorporating different cables and pipelines, can be a solution for integrating those independent utility services nowadays. These tunnels must be based on an endearing city planning and developed before extensive aboveground development.

While planning for the use of underground space, some design considerations must be taken into account in order to achieve a proper planning. Having an optimization between different aspects is important because it can ensure that the planning suits the process of urbanization of a city and maintains its sustainability.

2 CURRENT USE OF UNDERGROUND SPACE

2.1 *Underpass*

In developed city, aboveground transport is usually highly crowded, with insufficient space for pedestrian flow around the city. Also, aboveground walkway is easy suffered from adverse weather which make the condition worse.

Underground Pedestrian Street is built in many cities to handle this issue. The Underground Pedestrian Street contribute to sustainability by facilitate pedestrian to walk along the street to nearby area instead of using vehicles. In some cities, underpass is developed into underground shopping mall and tourist attraction. A summary of Underground Shopping Mall in Asia is shown below:

Table 1. Example of Underpass in Asia.

	Length (m)
Taipei City Mall (Taipei City Mall, 2016)	825
Osaka Namba Walk (Osaka Namba Walk, n.d.)	715
Fukuoka Tenjin Underground Shopping Mall (Tenjinchikagai, 2017)	590

2.2 Underground road tunnel network

To support the long-term economic growth with enhancing the access and connections to a city, large scale underground road networks are being planned or constructed in different cities. These underground road networks can provide extra capacity for through traffic and reducing the stain on local roads.

They can remove bottlenecks and relieve congestion for existing road network by provide a bypass for congested area. They can upgrade the city to a more livable city by reducing the traffic on surface roads, creating opportunities for urban renewal and improved public and active transport options.

With proper planning and design, they can also release new recreational green space for local communities. To reduce the visual impact and have a better control on the pollutant emission and noise emission, underground tunnel network becomes more popular although the construction cost is much higher than the service road network. With the advance in traffic control technology as well as the control of the tunnel ventilation system, long tunnel with various slip roads becomes more common to enhance the traffic flow. In some projects, long tunnels are interconnected or even constructed with slip tunnels which connected to underground carpark or bus terminus.

In Singapore, a 12km Kallang-Paya Lebar Expressway (KPE), which is Singapore's ninth expressway, was opened to traffic on 20 September 2008. The construction cost is US$1.2 billion (i.e. S$1.7 billion) and it includes a 9km of cut-and-cover underground tunnels with 6 traffic lanes. There are 8 interchanges and 6 ventilation buildings along the KPE. The Marina Coastal Expressway (MCE) is the tenth of Singapore's network of expressways. Construction for the MCE began in 2008 and was completed at the end of 2013. The 5km MCE connects with the southern end of the KPE and its junction with other road network. The MCE, with five lanes in each direction, handles the large number of commuters to be drawn to the offices, homes and recreational attractions there. MCE together with KPE forms the longest tunnel network in Singapore. They provide an additional high speed link to the city's heart to support the development of a New Downtown comprising Marina Bay Sands Integrated Resort, The Sail and Marina Bay Financial Centre, as well as other future developments (LTA, n.d.).

In Australia, WestConnex is part of an integrated transport plan to keep Sydney moving. It is a 33km predominately underground road network currently under construction in Sydney, New South Wales (NSW), Australia. The road tunnel network, a joint project of the New South Wales and Australian governments, encompasses widening and extension of the M4 Western Motorway (M4), a new section for the M5 Motorway (M5), and a new inner western bypass of the Sydney CBD connecting the M4 and M5. Together, these projects will create around 16km of new tunnels. In addition, 7.5km of the existing M4 will be widened and converted to a private tollway. To help fund the project, the publicly-owned M5 East Motorway (M5 East) will be converted to a private tollway, while the toll on the existing M5 will be extended for a further 34 years. The initial M4 widening and King Georges Road Interchange Upgrades began construction in 2015 and were completed in 2017. The M4 East and New M5

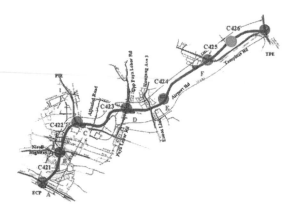

Figure 1.   Kallang-Paya Lebar Expressway (KPE) in Singapore (LTA, 2014).

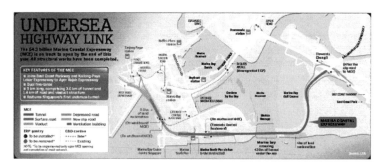

Figure 2.   Marina Coastal Expressway (MCE) in Singapore (Marina Coastal Expressway,2013).

tunnel stages started work in mid-2016 and are due for completion in 2019 and 2020 respectively. The final stages, the M4–M5 link, the Iron Cove Link and the Sydney Gateway are expected to be constructed between 2019 and 2023. Once land acquisitions, network extensions development costs and the cost of operations are accounted for, the total cost is forecast to be at least A$20 billion and possibly more than A$45 billion. The NSW Government has announced its intention to sell at least 51 per cent of WestConnex to yield between A$2 billion and A$4 billion (Wiki, 2018).

In Hong Kong, Route 6 is the newest underground road network under construction in Hong Kong. It will run through central Kowloon in Hong Kong and is marked as a high-priority trunk route in the Third Comprehensive Transport Study. The route starts off Route 3 near Yau Ma Tei, passes through the new Central Kowloon Route (CKR) and an unnamed submarine tunnel which together leads to the underground of Kai Tak Development Area, junctioning Route 5 on the way (Wiki,2018). CKR is a 4.7 km long dual 3-lane trunk road in Central Kowloon linking the Yau Ma Tei Interchange in West Kowloon with the road network in Kai Tak Development and Kowloon Bay in East Kowloon. The benefits of CKR include relieving congestion along the existing major east-west corridors, enhancing linkages between districts and underpinning various developments in Kowloon (HyD, 2012).

After the CRK, the route then junctions Kwun Tong Bypass of Route 2 in Kowloon Bay, and runs through Trunk Road T2 to Lam Tin, where it junctions with Route 2 again at the entrance to Eastern Harbour Crossing and runs into another proposed route, Tseung Kwan O - Lam Tin Tunnel (TKO-LTT) to southern Tseung Kwan O (CEDD, 2018). Trunk Road T2 is a dual two-lane trunk road of approximately 3km long with about 2.7 km of the trunk road is in the form of a tunnel. The TKO-LTT is

Figure 3. WestConnex in Australia (West Connex, n.d.).

a dual-two lane highway of approximately 4.2 km long, connecting Tseung Kwan O and East Kowloon (CEDD, 2018).

CKR together with the proposed Trunk Road T2 at KTD and Tseung Kwan O – Lam Tin Tunnel will form Route 6 with a total length of 12.5km that will directly link up West Kowloon and Tseung Kwan O. Route 6, when completed, is expected to relieve the congestion problem in Kowloon, and will also serve as an alternative route for the existing Tseung Kwan O Tunnel (Wiki, 2018).

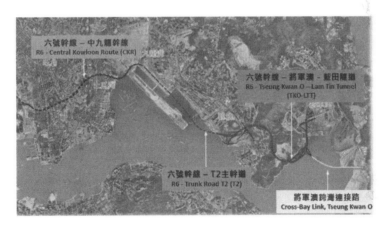

Figure 4. Route 6 in Hong Kong (Wiki,2018).

The challenges of having long underground road network includes (1) project cost, (2) control of airflow at tunnel junction during normal operation or fire emergency operation, (3) power supply for the plant, (4) air quality control, (5) noise control. In addition, there are operation and safety issues such as (1) fire and life safety, (2) flooding, (3) Electromagnetic Compatibility (EMC), (4) prevent claustrophobia, disorientation and tiredness for driving in a long underground road network (LUX, 2016).

2.3 *Underground metro systems*

A traditional usage of underground space in the cities is underground metro systems. It significantly reduces the use of vehicles within the city, lower the air pollutant induced by vehicles.

3 RECENT TREND ON THE USE OF UNDERGROUND SPACE

3.1 Combined services tunnels (Utility tunnel)

In some cities, utility services such as power cable, water pipework, gas pipework are directly buried underground independently. This method brings the following problems: (i) Lack of monitoring for the status of underground utilities, until it breakdown and services disruption occur. (ii) For maintenance, road excavation works would be needed, which cause disruption to traffic and produce noise and air pollution. (iii) As the pipework are buries underground by different parties, accidental damage of irrelevant services may occur and induce extra damage and inconvenience to the public.

For these problems, combined services tunnel, which already exist since 1850s, is a solution to these problems. By grouping various kind of utility services, such as telecom cables, water pipes and power cable inside with spaces allowed for access and maintenance. No excavation work would be needed to most maintenance work, and accidental damage to irrelevant utilities can be minimized.

By install sensors in the combined services tunnel, the condition of utilities can be closely monitored (e.g. flood sensor for water leakage, heat sensor for cable failure, gas detector for gas leakage/sewage pipe leakage), and maintenance can be planned before significant services failure.

In Singapore, combined services tunnel was built in Marina Bay, the tunnel contains cooling water pipe from district cooling plant, electrical cables and telecommunication cables, with space reservation for refuse conveyance system. The tunnel cost around US$ 1.26 billion, 1.4km long and with cross section of 22.7m width x 5m in height (China Trend Building Press, 2006).

3.2 Underground sewage treatment plant

The containing nature of underground space make it suitable for sewage water treatment plant or water treatment facilities, where odor can be contained within the cavern. However, the traditional wastewater treatment technology required large volume of tanks for biological reaction, which make Underground Sewage treatment hardly economical.

In Northern Europe, some wastewater treatment plants are located inside carven in order to reduce the impact of outdoor environment on the treatment process. In Stockholm (Verdict Media Limited, 2018), the expansion of Henriksdal Underground Wastewater Plant was developed in cavern. The expansion and renovation will enable the plant to treat 864 million litres of wastewater a day (i.e. $864,000m^3$/day).

Although the construction cost of a wastewater treatment plant is double if it is located underground instead of at grade in China, there is a trend to relocate the existing wastewater treatment plant to underground. This is because (1) it can create a better environment to the surrounding (e.g. better control on the noise and odor), (2) it can reduce the land use (i.e. smaller footprint) and (3) it can achieve a better use of resources as the plant can be located at city center (e.g. reduce the pipework and can reuse the treated water).

The challenges of having underground wastewater treatment plant includes (1) availability and cost of advance technologies that have small space with high efficiency, (2) power supply for the plant, (3) sludge handling. In addition, there are operation and safety issues such as (1) fire and life safety, (2) flooding, (3) explosion proof design for sensitive area, (4) leakage of chemical.

In China, with the use of Membrane BioReactor (MBR) or Moving Bed Biofilm Rector (MBBR), the spatial requirement for a large-scale wastewater treatment plant is in the order of 0.17~$0.4m^2$ per m^3 of wastewater. The construction cost is in the order of US$680~900 (RMB4500~6000) per m^3 of wastewater (污水处理观察, 2016).

In Hong Kong, Stanley Sewage Treatment Works, commissioned in 1995, is the first sewage treatment works built in caverns with a capacity of $11,600m^3$/day. Recently, a Shatin sewage treatment plant with a capacity of $340,000m^3$/day, is going to be relocated into a rock cavern. 13 hectares (i.e. 130,000m^2) of caverns, which is just half of the space of its original footprint, will be the largest sewage works relocated into a rock cavern in Asia. It will take around 11 years to complete the relocation. After the completion of the relocation, 28 hectares (i.e. 280,000m^2) of waterfront Shatin land can be freed up for housing or other community developments (Kao, 2017).

In Singapore, a large-scale desalination plant which can treat both seawater and freshwater will begin operations in 2020. The type of water that will be drawn into the desalination plant will depend on weather conditions. During wetter seasons, freshwater form the reservoir will be drawn. During dry seasons, the plant will draw seawater. The water drawn will then travel underground through a dual-flow chamber for around 1.8km. The water treatment plant will be located underground and capped with a rooftop lawn, which will be open to the public (Parcolnews, 2017).

Table 2.   Example of underground Wastewater Treatment Plant (污水处理观察, 2016).

	Plant	Country	Capacity	Process
1	Henriksdal Wastewater Treatment Plant	Sweden	864,000m^3/day	Membrane BioReactor (MBR)
2	Huaifang Water Recycling Plant, Fengtai District, Beijing	China	600,000m^3/day	MBR
3	Jinyang Underground Sewage Treatment Plant, Shanxi District, Beijing	China	320,000m^3/day (480,000m^3/day)	Anaerobic-Anoxic-Aerobic (AAO), MBR
4	Wenzhou Underground Sewage Treatment Plant, ZheJiang	China	400,000m^3/day	——
5	Viikinmäki Wastewater Treatment Plant	Finland	270,000m^3/day	Activated Sludge Process (ASP)
6	Geolide Wastewater Treatment Complex in Marseille	France	240,000m^3/day	ASP
7	Shenzhen Buji Sewage Treatment Plant	China	200,000m^3/day	ASP
8	Bishui Underground Wastewater Treatment Plant, Tongzhou District, Beijing	China	180,000m^3/day	ASP
9	Bekkelaget Water Treatment Plant	Norway	160,000m^3/day	ASP
10	Kunming No. 10 Sewage Treatment Plant	China	150,000m^3/day	MBR

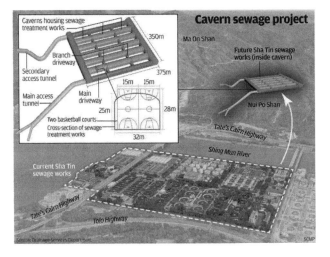

Figure 5.   Ambitious HK$30 billion project to vacate 28 hectares of land for housing or other community developments in Hong Kong (Kao, 2017).

Figure 6. Large-scale Desalination Plant in Singapore (Parcolnews, 2017).

3.3 Underground storm water storage/services reservoirs

In Singapore, there are lots of heavy rainstorms. It rains a lot and has approximately 2.4m (i.e. 94 inches) a year. There are rainwater collection system and rainwater storage system to avoid flooding. But no country has such a robust system already in place as Singapore, where half the land area is equipped to capture rainwater in gutters, barrels, tanks, and reservoirs. The most sophisticated of those systems is at the Changi Airport. Between 28 and 33% of all water used in the airport comes from captured rainwater, which is stored in two reservoirs. One reservoir balances the flow of water when tides are high, while the other collects runoffs from runways and green areas. Each year, the infrastructure saves the airport more than US$200,000 (i.e. S$275,000) for non-potable uses, like flushing toilets and performing firefighting drills (Weller, 2015).

Hong Kong's location puts it in the path of many tropical cyclones. Its annual rainfall is among the highest of cities around the Pacific. Urban flooding was once a serious threat. In Hong Kong, the overall flood control strategy has three parts. They are (1) interception upstream, (2) temporary underground storage midstream, and (3) upgrading pipes downstream. Hong Kong has used the concept of "sponge city" in a decade of infrastructure upgrades. The excessive rainwater is absorbed by large underground storage tanks during heavy rainstorms. The operation mechanism is based on real time water level monitored by sensors. Any rainfall that excess the handling capacity of the downstream drainage network will be temporarily stored in the underground water tank. For example, for the Happy Valley Stormwater Storage Scheme, 60,000m^3 of water, which is equivalent to water in 24 standard swimming pools can be stored. Other water is intercepted and diverted into the sea through massive drainage tunnels (Dan, 2016).

In mainland China, water-logging has plagued many cities during the heavy rainstorms season. A lot of cities (e.g. Beijing, Shanghai, Shenzhen, etc.) are exploring to construct "sponge cities" to solve the problem of urban flooding. Concrete surfaces have to be replaced by permeable materials and green spaces such as rain gardens to absorb and filter rainfall. Drainage systems have to be rebuilt to separate wastewater from rainwater. Huge underground rainwater storage tanks are being constructed so that rainwater can be collected effectively. The rainwater will be absorbed and purified, it will then be used for dust reduction on the road (i.e. cleaning the streets), greenbelt irrigation, car washing, firefighting, etc., if possible instead of discharge them to the sea or river. Large investment is required to achieve "sponge cities" in mainland China. It is expected that it will cost US$13~20/m^2 (i.e. RMB0.1~0.15billion/km^2) (Baidu wiki, 2018).

3.4 Waste transfer facilities/hazardous waste storage facilities

Underground space can also contribute to sustainable city by controlling unpleasant elements such as smell and noise. In solid waste transfer, the collection of residential waste comes with unpleasant smell, which disturbs the environment around the collection point and along the transport route.

Figure 7. Island West Transfer Station, Hong Kong (EPD, 2005).

Underground space can be utilized to reduce this situation by increase the capacity of waste collection point, which can lower the frequency of cleanup by garbage truck, saving operation cost and occasion for bad odor. Or utilize underground space to develop automatic pneumatic waste collection system to minimize the use of garbage truck to totally confine the air polluted (WSP, 2018).

In United States, Kissimmee was the first municipality to develop an underground refuse system. 17 mail-box like underground refuse bins were built above 11-foot-deep dumpsters in the ground. The dumpster can contain about 6 cubic yards of garbage. A truck was specially designed such that a crane is attached to empty the waste into the vehicle. US$500,000 was used to develop the project (Jackson, 2017).

In United Kingdom, a subterranean refuse system had been introduced in new district of the University of Cambridge. Steel bin chutes were set into the pavement that connects to large underground chambers. The system covers 150-hectare site and replaces 9,000 traditional bins 450 recycling and general waste bins. Collection of trash only takes place when the bins are full and trigger the sensor. The development took £1 billion (Ward, 2017).

In Hong Kong, a refuse transfer station is developed in an underground cavern in Kennedy Town. In the transfer station, refuse collection vehicle would unload the refuse into the compaction system, compact the waste into a tight container for transportation to landfill site by ship. The design capacity is 1,000 tonnes per day. To avoid impact to neighbour environment, waste water treatment plant and odour removal unit in exhaust system are installed (EPD, 2005).

4 MAJOR DESIGN CONSIDERATIONS

4.1 *Fire and life safety*

For underground development, fire and life safety is always one of the major concern. Since the underground development is confined, when fire occurs, it would be difficult to evacuate occupant inside and for fire fighter to pull out the fire. Moreover, the fire damage to the underground structure would be difficult, sometime impossible, to repair. Hence fire and life safety is one of the major concern for underground development.

4.1.1 *Fire compartmentation*
To hinder the fire spread to the entire development, proper fire compartmentation would be needed. Usually fire compartmentation can be done by fire rated wall and fire shutter. But for underground development with large cross section area, such as road tunnel and Underground Sewage Treatment Plant. There may be difficulties to install such items. In this case, drencher system could be considered.

4.1.2 *Fire detection*

For large underground development, traditional fire detection method, such as smoke detector and heat detector, are not sufficient to cover large area. With the latest technology, fire detection by Closed Circuit television (CCTVs) can be used.

4.1.3 *Automatic fire suppression system*

To minimize the fire damage, pull out the fire when it just starts is crucial. Automatic Fire Suppression System with fire detection system should be installed in underground development to pull out the fire at once. Currently, Automatic Fire Suppression System is installed in some underground metro stations and road tunnels, with experiment to prove their effectiveness in underground space (Steven et al, 2017).

4.1.4 *Escape route and egress route*

For underground development, escape route is highly limited. During planning stage, escape route should be carefully planned, with sufficient signage (e.g. lighting signage on ground) to indicate the route during fire.

Protected escape route should be arranged around the development for occupant to escape to relatively safer location quickly. For development that is deep underground that escape by foot is not possible, emergency escape vehicle (e.g. lifts, vehicles) should be arranged to facilitate the escape or arrange emergency capsule that can maintain life for few days should be arranged.

In Hong Kong University Station (Wiki, 2019), where the station is 100m below ground, vertical transportation is provided to facilitate evacuation process. In the station, 12 high speed lifts are used during normal operation and emergency with refuge area is provided as temporary rally point for people line up for the lifts. The refuge area is protected against fire and smoke by air pressurization and sprinkler systems. The lifts are protected by fire curtain and fire rated lift shaft, with 2 independence generators.

For adjacent underground development, sharing of fire escape route would be a potential way to lower the cost. However, demarcation line for management should be determined clearly to avoid area with no one manage.

4.2 *Impact to neighbor environment and communities*

To facilities sustainability of the city, consideration of impact of the underground development to aboveground environment and communities is important.

4.2.1 *Noise*

For some underground development (e.g. road tunnel, metro), noise pollution at portal or ventilation shaft would be significant. To avoid this, the location of noise source should far away from noise sensitive recipient (e.g. Hospital, school, residential area) if possible. On the other hand, noise absorption material, or noise barrier should be installed to lower the impact.

4.2.2 *Air*

For underground development, air pollution from underground development exhaust would impact the environment nearby. To tackle this, air purification systems could be adopted to remove harmful air pollutant like NOx and particulate from the air before discharge to aboveground, which had been adopted in Hong Kong Central Wanchai bypass (HyD, 2018).

Also, the upward exhaust shaft with high speed can be considered to dilute the air pollutant quickly while minimize the impact to nearby area.

If air pollution sensitive receiver (e.g. hospital, elementary school, elderly home) is nearby, long ventilation duct could be adopted to move the discharge point away from air pollution sensitive receiver.

4.3 Coordination with existing underground utilities

For planning, underground development, coordination with relevant parties for the location of underground utilities existing should be done to prevent impact to public by services disruption. Carefully planning should be done to allow re-routing of existing utilities, and through underground survey should be done to avoid damage of utilities not in record.

4.4 Coordination with city's master plan

To fully utilize the potential of underground space, coordination for underground space usage during city master planning stage is crucial. The construction sequence, usage of underground area should be coordinated to optimize the excavation work and avoid crash of underground development. Moreover, the site boundary and the party responsible should be carefully demarcated to ensure all underground development is well managed.

For metro system, in the development stage of city, metro station can be built in planned development area before actual development, to minimize the impact of metro construction work to traffic and other development works when those area is under development. In Singapore Circle Line, Bukit Brown station is built in undeveloped area at Novena planning area, reserved for future development of the area (Wiki, 2018). In Hong Kong, non-operational Kwu Tong Station is built in rural Kwu Tong Area at 2018, reserved for future development of interchange station of Northern Link and East Rail Line for North East New Territories Development (Wiki, 2018).

In the overall city planning, crowd control would be one of concern for congested area, especially around metro station and Central Business District. One of the solution is to identify potential congestion point by simulation, and planning ample walkways or sharing walkways of metro station and nearby shopping mall.

5 CONCLUSION

In the last few decades, underground space has become increasingly important for society's development. Apart from underground metro system and road tunnel network, fully utilized the underground spaces can contribute to the sustainability of a city. It can create extra space for the further growth of the city. Also, at grade existing facilities such as wastewater treatment plant, water treatment plant, waste transfer and storage facilities, etc. can be relocated in underground space or cavern in order to release the at grade space. At the same time, the noise and the odor from the facilities can be confined and treated to reduce the impact to the surrounding and hence increase the land value as well as the living condition of the surrounding area. To achieve a cost-effective use of the underground space, engineers have to work with the city planners to establish a Master Planning on the development of the underground space. Sharing of resources and user-friendly facilities can be integrated into the Master Plan. If necessary, advance work for some facilities will be constructed during the construction of other facilities so as to avoid costly construction and disturbance to others in the latter stage.

REFERENCE

China Trend Building Press, 2006. Common Services Tunnel and district cooling system. Construction & Contract News 2006 No.3 From http://www.building.com.hk/forum/2007_0309marinabay.pdf
Chris Weller, 2015. Singapore has come up with an ingenious way to save water. Business Insider.
Civil Engineering and Development Department (CEDD), HKSAR., 2018. Kai Tak Development - Trunk Road T2. Retrieved from http://www.trunkroadt2.gov.hk/en/overview.html
Civil Engineering and Development Department (CEDD), HKSAR., 2018. Tseung Kwan O - Lam Tin Tunnel. Retrieved from http://tko-ltt.hk/en/
Environmental Protection Department (EPD), 2005. Island West Transfer Station. Retrieved from https://www.epd.gov.hk/epd/english/environmentinhk/waste/prob_solutions/msw_iwts.html

Ernest Kao, 2017. Hong Kong sewage plant to move into caverns in 11-year plan. South China Morning Post, 20 Novemember 2018. Retrieved from https://www.scmp.com/news/hong-kong/health-environment/article/2120573/hong-kong-sewage-plant-move-caverns-11-year-plan

Henriksdal Wastewater Treatment Plant, Stockholm. Water Technology. Retrieved from https://www.water-technology.net/projects/henriksdal-wastewater-treatment-plant-stockholm/

Highways Department (HyD), HKSAR., 2012. Central Kowloon Route. Retrieved from http://www.ckr-hyd.hk/en/about.php?page=2

Highways Department (HyD), HKSAR., 2018. Central WanChai Bypass. Retrieved from http://www.cwb-hyd.hk/en/

Ken Jackson, 2017. Kissimmee first in the nation to use underground trash system. Osceola News-Gazette, 20 April 2017. Retrieved from http://www.aroundosceola.com/news/business/kissimmee-first-in-the-nation-to-use-underground-trash-system/article_f4b5923f-c7f4-5e7c-8ac4-9d70ae97f02b.html

Lands Transport Authority (LTA), Singapore. Marina Coastal Expressway. Retrieved from https://www.lta.gov.sg/content/dam/ltaweb/corp/RoadsMotoring/files/MCE%20Brochures.pdf

Lands Transport Authority (LTA), Singapore., 2014. Kallang-Paya Lebar Expressway (KPE). Retrieved from https://www.lta.gov.sg/content/ltaweb/en/roads-and-motoring/projects/kallang-paya-lebar-expressway-kpe.html

LUX, 2016. Surreal' lighting to make 'world's longest tunnel' safe. Retrieved from http://luxreview.com/article/2016/08/-surreal-lighting-to-make-world-s-longest-tunnel-safe

Mao Dan, 2016. Hong Kong becomes 'sponge city' to stop floods. CCTV.com Retrieved from http://english.cctv.com/2016/09/26/VIDEcDmWnkHg1d1a5KISBETt160926.shtml

MCE Tunnel. Retrieved from https://cdn.runsociety.com/wp-content/uploads/2013/12/17232152/mceSG.jpg

Osaka Namba Walk. Retrieved from https://walk.osaka-chikagai.jp/en/

Parcolnews, 2017. World's First Large-scale Desalination Plant in Singapore. Retrieved from http://www.parcolnews.com/2017/08/worlds-first-large-scale-desalination-plant-singapore/

Sponge City, Baidu wikipedia. Retrieved from https://baike.baidu.com/item/%E6%B5%B7%E7%BB%B5%E5%9F%8E%E5%B8%82/16012711

Steven Lai, Tim Cho & William Xie, 2016. Use of Fire Suppression System in Underground Stations and Metro Tunnels.

Taipei City Mall (2016). Retrieved from http://www.taipeimall.com.tw/zh-TW

Tenjinchikagai, 2016. Fukuoka Tenjin Underground Shopping Mall. Retrieved from http://www.tenchika.com/

Victoria Ward, 2017. End of the wheelie bin in sight as it is replaced with underground system. Telegrap, 12 September 2017.

WestConnex. Retrieved from https://www.westconnex.com.au/

Wikipedia contributors. "Bukit Brown MRT station." In Wikipedia, The Free Encyclopedia. Retrieved from https://en.wikipedia.org/w/index.php?title=Bukit_Brown_MRT_station&oldid=868799978

Wikipedia contributors. "HKU Station" In Wikipedia, The Free Encyclopedia. Retrieved from https://en.wikipedia.org/w/index.php?title=Special:CiteThisPage&page=HKU_station&id=858636014

Wikipedia contributors. "Kwu Tung station." In Wikipedia, The Free Encyclopedia. Retrieved from https://en.wikipedia.org/w/index.php?title=Kwu_Tung_station&oldid=854841220

Wikipedia contributors. "Route 6 (Hong Kong)", In Wikipedia, The Free Encyclopedia. Retrieved from https://en.wikipedia.org/w/index.php?title=Route_6_(Hong_Kong)&oldid=857063638

Wikipedia contributors. "WestConnex." In Wikipedia, The Free Encyclopedia. Retrieved from https://en.wikipedia.org/wiki/WestConnex

WSP., 2018. Underground Space Utilization

污水处理观察, 2016. 大型地下式污水处理厂的中国身影

简健文, 2015. 地下式污水處理廠的特點與分析. 建築工程技術與設計.2015年第三期

Tunnels and Underground Cities: Engineering and Innovation meet Archaeology,
Architecture and Art, Volume 10: Strategic use of underground
space for resilient cities – Peila, Viggiani & Celestino (Eds)
© 2020 Taylor & Francis Group, London, ISBN 978-0-367-46878-1

Noisy-Champs station: Coordinating underground, design and engineering

D.S. Da Silva Leite

SYSTRA, Association Française des Tunnels et de l'Espace Souterrain (AFTES), Université Paris-Est Marne-la-Vallée (UPEM), Paris, France

ABSTRACT: Designed as one of the major entrances to the new Greater Paris subway system (the Grand Paris Express), Noisy-Champs Station has become one of the major flagship projects. Noisy Champs will be an interchange station between the Lines 15 & 16 (two new rapid transit lines from GPE network) and an existing suburban railway line (the RER A line) with a daily traffic of 150,000 people and serving a large surrounding population. Designed by Jean-Marie Duthilleul, this major hub can be explained as an open garden leading passenger down through its space to the platforms. Passengers enjoy this special urban landscape from their arrival until their departure. Noisy-Champs is one of the most significant current projects in Paris' suburban Ile de France region, a project where the greatest designers, builders and coordinators are working hand in hand to ensure the realization of this new symbolic gateway to Paris

1 DESCRIPTION OF THE PROJECT

The Nouveau Grand Paris project, announced on 6 March 2013, aims to modernize the existing transport network and build a new automated metro, the Grand Paris Express (GPE). The rollout and financing of the Grand Paris Express is managed by the Société du Grand Paris, a government-created public company. The Société du Grand Paris is responsible for building the infrastructures required for the new metro system and purchasing the required rolling stock. It also carries out development operations in areas impacted by the Grand Paris Express. This project comprises the following lines:

Line 15: new high-capacity loop line around Paris, offering a maximum number of connections with the existing transport networks within the dense urban area;

Lines 16, 17 and 18: new metro lines of suitable capacity to serve emerging areas (east of Seine-Saint-Denis, Grand Roissy, South-West Île-de-France);

Lines 14 and 11: extensions to the existing metro lines. Line 14 is extended to the north (as far as Saint-Denis Pleyel) and to the south (as far as Orly airport), and line 11 to the east, from Rosny-Bois Perrier to Noisy-Champs.

Line 15 (SGP 2017) is an underground metro rail line approximately 75 km long, passing through three departments of the inner suburbs and connecting to line 16 to the west of Seine-et-Marne, in the shared terminus of Noisy-Champs. Trains are planned to run every 120 seconds in peak periods and the journey time for the whole line will be 80 minutes.

The project has been divided into sections according to a construction schedule, starting with the southern section of Line 15 (Pont-de-Sèvres to Noisy-Champs).

The southern section is divided into two packages: T3 and T2. Package T3 runs from Pont-de-Sèvres to Villejuif Louis Aragon (including station) and passes through the Hauts de Seine and Val-de-Marne departments with eight stations. Package T2, which includes the Noisy-Champs section, extends 21 km from Villejuif Louis Aragon station (not including the station) to Noisy-Champs (including the station). This fully underground package mainly runs

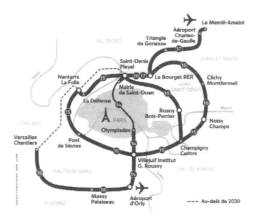

Figure 1. The Grand Paris Express Project. Source: Société du Grand Paris.

Figure 2. The southern section of line 15. Source: Société du Grand Paris.

through dense urban areas and includes three under-river crossings: one under the Seine and two under the Marne.

It serves eight stations connecting with planned or existing urban and main rail transport lines.

All the stations are within the Val-de-Marne department, except for the Noisy-Champs site, which has the distinction of being on the border of two departments – Seine-Saint-Denis and Seine-et-Marne. This section includes an infrastructure maintenance depot located at Vitry-sur-Seine, and a train maintenance and storage depot at Champigny-sur-Marne.

2 DESCRIPTION OF THE NOISY-CHAMPS SECTION

The Noisy-Champs section is comprised of a set of works with a total length of around one kilometre, spanning two towns and two departments: Noisy-Le-Grand (department 93 – Seine-Saint-Denis) and Champs-sur-Marne (department 77 – Seine-et-Marne).

It comprises the station and underground rail infrastructures in the north and south and passes through wooded, sparsely-inhabited areas, formerly reserved for the construction of motorway A103, to the eastern side of the Boulevard du Rû de Nesle.

The central element of this segment is the future GPE Noisy-Champs station, built beneath and around the tracks and platforms of the existing station of the same name on line A of the RER (regional express network). The station is situated between the road bridge on the Boulevard du Rû de Nesle and the current passenger building of the RER station.

Figure 3. The Noisy-Champs section. Source: SYSTRA.

The line 15 rear station is situated on the north of the station. It includes the line 16 junction structure, the cut-and-cover tunnel for line 15 and the 0801P ancillary structure. This area is located entirely within the municipality of Champs-sur-Marne.

The line 15 front station is situated on the south of the station. This area includes the line 16 rear station, the line 15 front station, the lift tracks linking the two lines and, at its southern end, the 0802P ancillary structure. The front station structures are located entirely within the municipality of Noisy-le-Grand.

3 CONTEXT AND CHOISE OF INFRASTRUCTURE

The geological context of this section is mainly a layer of fill, a layer of green clay approximately 7 metres deep, two layers of marl (Argenteuil marl and Pantin marl) approximately 20 metres deep and a limestone layer (Champigny limestone) approximately 30 metres deep.

The whole of the Noisy-Champs segment is open-air and follows a phased construction schedule: diaphragm walls, raft (level L16), intermediate slabs (slab level L15 and upper slab level L15) and cover slabs. The average depth of the works is 25 metres and the diaphragm walls are set into the limestone at an average depth of 35 metres.

4 NOISY-CHAMPS STATION

The landmark station of Noisy-Champs on the Grand Paris Express network is situated in an urban and fast-changing rural setting.

Located at the eastern terminus of line 15, it is a major interchange hub for the metropolitan area, offering a large number of connections and destinations. It lies at the heart of an intermodal system within a public space that gives priority to pedestrians, buses and soft transport. This system connects up all the above-ground transport modes, the three metro lines 15, 16 and 11 and the existing RER-A line.

The station is the central element of an ambitious urban development plan to create a new quarter around a transport interchange, serving as a stimulus for change in the neighboring suburbs. The project redefines the Boulevard Rû de Nesle area, promoting the expansion of new urban facilities in the neighboring areas and opening up the existing outlying districts of the municipalities.

Jean-Marie Duthilleul, the designing architect, has sited the station in the middle of a garden square. The station, following the line of the future Boulevard du Rû de Nesle, will serve as a landmark in the urban landscape. Its presence is a strong signal, a unifying feature providing this meeting point with an identity.

As the site will be developed in successive stages over a number of years, the station has been designed to accommodate each of the development stages:

Line 15: Within the immediate environment, which will remain largely unchanged, the project connects to the existing infrastructure, including the bridge and current Boulevard du Rû de Nesle, and the ends of Newton and Archimède boulevards. A second bridge to the east, over the RER-A line, will be added for continuity of road access. The land above the underground infrastructure to the north and south, over the pre- and post-arrival tracks, is earmarked for future development.

Line 11: Following work carried out in the same context, having little impact on station operations, the terminus for line 11 will be opened and connected to Noisy-Champs station.

Urban project: A frontage will be constructed around the station square and the new Boulevard du Rû de Nesle opened, restructured to the east of its current line and following the north-south axis of the station. The areas alongside the boulevard will be built. A main pedestrian path will run to the west, above the partially covered RER-A. The areas around the station will also be restructured.

Noisy-Champs station has been designed as an "open-air" station beneath the level of the surrounding urban space. A major challenge is to preserve the continuity of the public spaces, from ground level to the platforms for all the metro and RER lines. This continuity will be provided by fluid circulation paths, open views, continuous natural light and by keeping a consistent ambiance between the various levels.

The site is made up of three areas, each with a specific function: intermodality, representation and architectural presence, but also urban transition and flow.

The station concourse is located centrally, giving direct access to all the metro lines and to the RER-A in both directions. It lies cross-wise, following the main pedestrian route through Noisy le Grand and Champs sur Marne, leading to the towns to the east and west.

The northern and southern entrances to the station are from two broad piazzas to the north and south of the central concourse parallel to Archimède and Newton boulevards. This intermodal hub provides connections between the metro lines, buses and car and bicycle parks.

These piazzas are bordered by two planted patios. These bring natural light and greenery as far as the platforms, providing continuity of the urban landscape. They serve both a technical and a security purpose and provide access to the car parks, thus avoiding any high-level exits onto the piazza.

Inside, the layout of the connections is as simple and logical as possible in a cross-shaped infrastructure.

The underground metro lines of the Grand Paris Express, one above the other, terminating from the south for line 15 and from the north for line 16, intersect with the open-air tracks of RER-A running east to west.

The height of the RER-A platforms and the thickness of the supporting structure for the tracks (3.40 m) dictate the exact position of the underground tracks for metro lines 15 and 16 and their respective platforms.

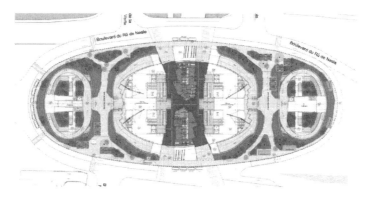

Figure 4.   Noisy-Champs concourse level. Source: Société du Grand Paris/AREP/Duthilleul.

The station has been organized to respond to the following essential objectives: to enhance the visibility of functions and pathways, to provide a visual link between the station and its surrounding urban spaces, to facilitate access to transport and intermodality, to provide natural continuity between transport and the town and to bring natural light to the whole station and safety and security to the transport spaces.

4.1 Description of the station

The four levels of Noisy-Champs intermodal station correspond to a particular mode of transport and serve a precise function, from the piazza to the platforms:

The concourse level (+97.10 NGF) is the main reception space for shops and services from town level. It provides access to all the modes of transport that pass through the station, including the RER-A in both directions;

Figure 5.   Noisy-Champs station concourse. Source: Société du Grand Paris/AREP/Duthilleul.

Figure 6.   Noisy-Champs station line 15 platform level. Source: Société du Grand Paris/AREP/Duthilleul.

The platform level of RER-A (+91.40 NGF) will initially receive and distribute flows between modes and connections to RER-A and metro lines 15 and 16. Various technical rooms, such as Lighting Power Stations (LPS), HVAC rooms, communications voltage and low-voltage plant rooms will also be located here. In future, this level will also serve the parking areas and the connection with metro line 11;

The L15 platform level (+83.90 NGF) will have connections with line 16 and emergency exits leading to the open gardens. The technical rooms are also located at this level, including some of the staff rooms, ventilation rooms and communications voltage/low-voltage plant rooms;

The L16 platform level (+76.15 NGF) will have on north and south sides the staff rooms for lines 15 and 16, the automated control rooms, the main communications voltage plant rooms and tunnel decompression plant rooms. The emergency exits to the gardens are also located at the ends.

5 INTERFACE WITH THE RATP

The station is sited at right angles to the existing RATP lines for RER-A. RATP first has to construct a supporting deck for the RER-A tracks at the future GPE station site. This 'deck' type frame measuring approximately 43 x 21 m comprises a raft level accommodating the RER-A platforms and a cover slab which will partially serve as the slab for the station entrance hall. This structure will be built to the south of the existing tracks and moved into place over a long weekend of a planned stoppage to RER operation in November 2018.

A gantry at platform level will be used to the connection with the walkways, the car parking areas and other station levels.

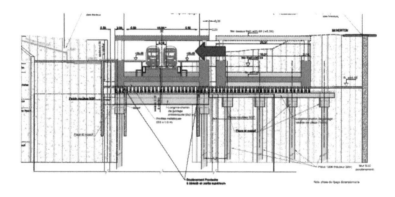

Figure 7. The RATP "deck" type structure. Source: Société du Grand Paris.

6 DESCRIPTION OF UNDERGROUND RAILWAY STRUCTURES AROUND THE STATION

6.1 *Rear station (0801P ancillary structure, cut-and-cover tunnel, line 16 junction)*

The Noisy-Champs rear station on the north involves a 460-metre cut-and-cover tunnel structure divided into three parts:

A 144-metre-long section, varying between 20 and 42 metres in width, providing a branch-off for line 16 from line 15 cut-and-cover tunnel structure. The eastern zone of this structure is 65 m long and in the preliminary phase provides the launch area for the line 16 tunnel boring machine. As a final stage, it will include various plant and technical rooms. This structure is composed from the raft:

The L16 level, at a depth of approximately 22 metres, including the tracks and some of the plant rooms for line 16, such as the rectifier room;

The L15 level, located around 13 metres deep, including the first part of the terminus storage sidings for line 15 and, in the eastern part, the plant rooms for lines 15 and 16 (high-voltage distribution station, switching station for line 15 and Lighting Power Station for line 16);

The car parking level around 7 metres deep, which includes a car parking area and the parking technical rooms;

A cut-and-cover tunnel for line 15 located in the central part with 308 metres long. It is a cut-and-cover structure on a single level 18.40 metres wide and includes the storage sidings for the line 15 terminus.

The 0801P ancillary structure is situated in the northernmost zone. It will serve as the emergency access shaft and provide ventilation and smoke extraction technical rooms plants. Measuring 48 metres long and 20 metres wide, it is composed from the raft:

The L15 level at approximately 15 metres deep, that includes the ventilation and smoke extraction room for line 15, the pumping station and an evacuation access The L15 level at approximately 15 metres deep, that includes the ventilation and smoke extraction room for line 15, the pumping station and an evacuation access;

The intermediate level at approximately 7 metres deep, includes another part of the ventilation and smoke extraction room for line 15, the Lighting Power Station, the communications voltage and HVAC plant rooms.

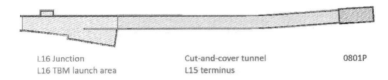

L16 Junction   Cut-and-cover tunnel   0801P
L16 TBM launch area   L15 terminus

Figure 8.   Noisy-Champs rear station. Source: SYSTRA.

6.2   Front station (lift tracks, backup control center, 0802P ancillary structure)

On the south area of the section, the Noisy-Champs front station is a cut-and-cover tunnel structure 350 m long and varying width between 22.45 and 11.70 metres, composed by:

At the southern end, the 0802P ancillary structure 94.80 m long composed by the line 15 TBM launch area in the preliminary phase. This structure will be an emergency access shaft and it will provide tunnel ventilation and smoke removal in the final phase. It will be composed from the raft;

The L16 level approximately 26 metres deep, including the buffers stop for line 16 and the pumping station;

The L15 level approximately 18 metres deep, including the L15 tracks and an evacuation access;

The L15 upper level at approximately 9 metres deep, including some of the technical rooms for lines 15 and 16, such as the tunnel ventilation and smoke extraction room and a switching station;

Between 0802P ancillary structure and the station, the front station is a cut-and-cover tunnel structure 255 m long, comprising a parking level (also referred to as the upper level of L15) and the L15 and L16 levels. This structure includes at both sides a 4.5-metre-wide lift tracks that connect the levels of lines 15 and 16 with a 3% gradient, composed from the raft:

The L16 level approximately 23 metres deep, includes the lowest point of the lift tracks and the stabling areas for the line 16 terminus.

The L15 level approximately 16 metres deep, includes the line 15 tracks and the top of the lift tracks;

The upper level of line 15, approximately 8 metres deep, includes another part of the technical areas for lines 15 and 16, such as ventilation for line 15, ventilation for line 16, backup control center, rectifier station for line 15 and switching rooms for lines 15 and 16, used either for power to the tracks or for the lift tracks.

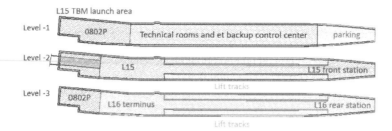

Figure 9.   Noisy-Champs front station. Source: SYSTRA.

7   WORKS

The Noisy-Champs section is divided into three civil works packages, mainly relating to the planning requirements for interfaces with other operators and the launch of the TBMs.

To the north, the rear station area (L16 junction structure, cut-and-cover tunnel and 0801P ancillary structure) are in package T2E. This is the first to be built because a part of the structure will be used for the future line 16.

To the south, the 0802P ancillary structure belongs to package T2C. This package comprises the tunnel between Bry-Villiers-Champigny station and 0802P ancillary structure, and all other ancillary structures along this tunnel. This tunnel will be built by the TBM launched from 0802P, which is also a planning priority. This package also includes 0807 ancillary structure that is the junction structure connecting the main tunnel to the future maintenance and storage depot.

The Noisy-Champs station and front station correspond to package T2D, which is the last civil engineering contract for the sector to have been announced; the station works will start once the RATP supporting deck are in its final place.

7.1   TBMs

7.1.1   Line 16 junction structure and TBM
The L16 junction structure located on Noisy-Champs rear station is used as a TBM launch area for this line. Once the TBM's launch operations and the works will be completed, the plant rooms for line 16 will occupy this strategic zone due to its proximity to the station.

Figure 10.   The line 16 TBM launching area. Source: SYSTRA.

7.1.2   0802P and line 15 TBM
The 0802P ancillary structure located at the front station will be used as the line L15 TBM launch area. Once the TBM's launch operations and the works will be completed, the plant rooms for line 15 will occupy this space.

Figure 11. The line 15 TBM. Source: SYSTRA.

8 BIM AND TECHNICAL COORDINATION

In view of the intermodal nature of this project characterized by the large number of inter-faces, Building Information Modelling BIM (EGF BTP 2015) is used to improve coordination between the contractors and to optimize management of the project and the works on the Noisy-Champs sector. By using BIM and the 3D models, the coordinator has real time access to changes and updates to the project made by all the contractors. Thus, consistent and reli-able data are available all throughout the project and its construction. A team of planners, BIM managers and coordinators are working every day across the different disciplines involved in the project (civil engineering, planning, architecture, MEP and communications voltage, track systems, construction), without interruption to production and planning. There-fore, the sector coordinator always has up-to-date information, making it possible to antici-pate any problems regarding interfacing, synthesis and impact on planning and costs, and facilitating all exchanges with the local authorities concerned, the client and the work contrac-tors. The coordinator also brings all the elements of the project together, using the contrac-tors' own models. These 3D models are a reliable means of real time monitoring progress thus resolving any problems as they arise; this is an efficient way of achieving a sound and consist-ent project from the initial design stages right through to its completion.

Figure 12. Example of BIM coordinating model of Noisy-Champs station. Source: SYSTRA.

9 CONCLUSION

Acclaimed as the new eastern gateway to Paris, Noisy-Champs station on the Grand Paris Express network demonstrates, by the complexity of the project and its realization, the importance of creating an intermodal hub for the public transport network in the Paris region. The significance for the whole area goes beyond just the building of a station; the project will have a lasting impact on two towns and two departments and on the whole of the metropolitan area and its inhabitants, improving considerably their quality of life. The size of this project also demonstrates the complexity of this type of infrastructure, especially when connected to an already dense and saturated network. The large number of contractors from all disciplines are working towards a common goal – to create a landmark station for Grand Paris Express, showcasing the very best of the skills and expertise of the engineers, designers and architects in Paris and in France.

REFERENCES

Entreprises Généreales de France BTP 2015. Note de position: *Intégrer le processus BIM dans un marché public globale* Paris: Le Moniteur
Société du Grand Paris 2017. Dossier de presse: *La Ligne 15 sud la première ligne du Grand Paris Express* Paris

*Tunnels and Underground Cities: Engineering and Innovation meet Archaeology,
Architecture and Art, Volume 10: Strategic use of underground
space for resilient cities – Peila, Viggiani & Celestino (Eds)*
© 2020 Taylor & Francis Group, London, ISBN 978-0-367-46878-1

Managing knowledge for future-proof tunnels in The Netherlands

K. de Haas
*COB, Netherlands Knowledge Centre for Underground Space and Underground Construction,
Delft, The Netherlands*

W. Broere
Delft University of Technology, Delft, The Netherlands

H. Dekker
Rijkswaterstaat, Ministry of Infrastructure and Water Management, Den Haag, The Netherlands

L. van Gelder
Soltegro, Capelle aan den IJssel, The Netherlands

ABSTRACT: The COB is The Netherlands Knowledge Centre for Underground Construction and Underground Space. Over sixty organisations, from government, industry and knowledge institutes, work together to learn and develop smart solutions regarding tunnels. This network has developed a long-term vision on tunnels in The Netherlands. As a result, nineteen tunnels and all stakeholders collaborate on eight research topics of our tunnel programme. We found common ground and the key issues are already being addressed in both practice and science. Topics include risks, construction failure, modular renovation, digital tunnel twin, virtual testing, virtual education and system safety. The way clients, science and industry work together within the COB network has been a major precondition to achieve these great results in such a short period of time.

1 INTRODUCTION

In December 2016 COB, The Netherlands Knowledge Centre for Underground Construction and Underground Space, published a long-term vision on tunnels in The Netherlands (COB, 2016). In 2017, two key tasks were distilled from this vision document: how to create additional value for society and how to reduce disruptions, during normal operation, in emergency situations, as well as during maintenance and renovation works. In 2018, the COB network drafted four development lines in a tunnel programme (COB, 2017a) from these key tasks: civil construction and renovation, the digital tunnel twin, tunnels and their value for their surroundings and adaptive installations. The first two development lines have already resulted in detailed work plans for 2018 to 2021. The detailing of the other two development lines will start in 2019.

This paper deals with the two development lines already prepared, each of which is divided into several projects. It includes a summary of the detailed projects, goals and deliverables for the coming years.

Figure 1. The COB has created an online overview of Dutch tunnels. It shows that twenty-eight tunnel projects (green = new, orange = renovation) have been scheduled for the upcoming ten years.

2 DEVELOPMENT LINE 1: DIFFERENT CIVIL CONSTRUCTION AND RENOVATION

At least twenty road and rail tunnels are scheduled to be renovated in The Netherlands in the next ten years. Many of these tunnels have been constructed in the 1960s and 1970s as immersed tunnels and half their expected life-span has now passed (Bijl, 2011). Determining the scope of (civil) renovation of these tunnels is a complicated task and, as a long-term full closure of the tunnels would lead to severe public disruptions, modular renovation is becoming increasingly necessary. Changes in the scope of renovation works after contracts have been awarded, or tunnel failures due to unexpected and unscheduled maintenance during operation are undesirable for all parties.

Because of the high traffic intensities in the densely populated Netherlands and the limited available budgets, renovating tunnels is a far-reaching process with many administrative-political considerations. The complete closure of tunnels for a longer period is usually not possible because of the essential function of tunnels in the Dutch road network. Very careful planning of smart (innovative) renovations has therefore become essential.

This development line focuses on four goals:

1. Being able to better determine the complete civil scope of tunnel renovations.
2. Being able to apply predictable special and extensive maintenance works, integrated as part of the regular management and maintenance systems of tunnels.
3. Reduce the (social) nuisance (public disruption) by reducing the amount of time required for (major) maintenance and renovation with associated reduced availability of the tunnel to traffic.
4. Reduce the occurrence of major renovation works by addressing 1, 2 and 3. Eventually eighty per cent of the (civil) maintenance of tunnels will have to take place as part of regular maintenance cycles. We thus achieve one of the main objectives of the tunnel programme, which is less disruption due to higher availability to end users and predictability of maintenance and availability.

These four goals have been translated into four projects:

1. Identify the risks (contributes to all goals).
2. Identify and reduce knowledge gaps regarding structural failure (contributes to all goals).
3. Modular renovation, blueprint and assessment framework (contributes to goals 3 and 4).
4. 'Know your tunnel', framework for a handbook and manual for a specific tunnel (contributes to all goals).

2.1 Project 1: Identify the risks

Regular inspections and preparatory research prior to a tender for renovation, are not always sufficiently substantive to really identify all the detailed aspects of the tunnel construction and its current state. After being awarded the job, the contractor usually carries out additional research, followed by a discussion about scope extensions or, in the worst case, changes to the schedule.

2.1.1 Project goal and deliverables

In this project, we want to conduct at least twenty interviews based on an extensive questionnaire to identify what experiences have been gained (and are being gained) in scheduled, ongoing and completed renovation projects. We also want to make use of the lessons learned from the new Rijkswaterstaat inspection programme. All this information will result in a risk checklist for tunnel renovation, management and maintenance. This checklist should become an important tool for (long-term) programming, scope assessment of renovations and planning of regular large and small maintenance works.

This risk checklist must include:

A. The most common risks
B. The associated failure mechanisms.
C. The associated inspection techniques, with a distinction between effectiveness, impact on the availability, level of predictability, and amount of damage to tunnel integrity.

It is expected that this stocktaking will identify several risks for which we do not yet know the failure mechanisms and for which there are no inspection techniques yet. These knowledge gaps will be passed on to the second COB project: 'Structural failure'.

2.2 Project 2: Structural failure

The frequent unexpected and unscheduled reduced availability of bridges, locks and tunnels in recent years has shown that we still have gaps in our knowledge regarding (residual) lifespan, failure mechanisms and the effectiveness of control measures. For too long it has been thought that civil structures are hardly subject to ageing. More and more often, we are confronted with joint problems due to leakages, subsidence and rotations of tunnel elements and even failure of components such as tension piles, reinforcement bars and fire resistance plating. It cannot be ruled out that other unexpected failures may occur in the future.

This project will provide more fundamental knowledge of structural failure mechanisms so that unexpected failures can be kept to the absolute minimum. Determining the true lifespan of civil structures is the most fundamental knowledge goal for this project. Determining the true-life-span and the rate of ageing of structures is an issue faced by many projects worldwide, and a lot of research is being conducted in this field all around the world. We will incorporate the most recent research results when they become available.

This project foresees an approach in which the contribution of tunnel project managers and technical staff, academia and industry experts is integrated by one central steering committee on a permanent basis, like a spider in the web, below which, for the time being, three subcommittees are set up.

In addition to coordinating and connecting activities to ensure coherence in the different COB projects, the steering committee will work on a 'structural health report' and 'structural health monitoring' as tools for predictable maintenance. These instruments are used to answer questions such as: if a tunnel is subject to subsidence, how will the tunnel react, and do we need to take measures inside or outside the tunnel, or do we possibly need to apply soil improvement techniques below the tunnel?

Other knowledge gaps already identified are related to immersion joints, omega profiles, tension piles, permanent on-line monitoring systems, new monitoring techniques using sensors, inspection techniques and determining the residual lifespan of tunnels and their components.

2.2.1 *Three subcommittees*

1. Joints

The objective of this committee is to understand the failure mechanisms of a specific civil part of tunnels: the immersion joints that provide the primary flood defence in immersed tunnels. The committee put together an expert team that advises ongoing tunnel projects and other stakeholders on the inspections, control measures, and so on. that need to be carried out. Based on questions arising from the Eerste Coentunnel renovation project, this work was continued with the Heinenoordtunnel and the Kiltunnel. The experience already gained has been translated into knowledge and experience in guidelines.

After a detailed analysis of the problems associated with immersion joints, the committee has now entered the next stage. The committee's most important activities will include:

A. Advise Rijkswaterstaat, as the national road and tunnel operator, as well as the owner of the private Kiltunnel on the scope of the studies for renovation of the Eerste Heinenoordtunnel and the Kiltunnel and critically assess the results after implementation. These two tunnels are quite similar in construction details, geological conditions and age, and are expected to face quite similar challenges during renovation.

B. Retain knowledge and making it available to future renovation projects. As an example, the results of endoscopic investigations into immersion joints, to obtain a detailed assessment of the current state of the joints, rubber seals and bolts, will be made public through the COB knowledge bank, as will the action plans for the Heinenoordtunnel and the Kiltunnel renovations.

C. Investigate possible problems of all types of joints, not only immersion joints but also closure joints between concrete sections and dilation joints. This comment will be the centre of knowledge regarding joint investigation and renovation in immerged tunnels.

2. Deformation of tunnels

A second important issue that governs the overall behaviour of the civil structure is the deformation of tunnels. How do the structure and the soil below it behave, what are the associated failure mechanisms and which control measures are required and effective? At the Kiltunnel project, the Delft University of Technology together with the University of Zagreb experimented with a MASW soil scan that, in combination with a 3D model of the tunnel itself and including time dependent soil behaviour, should result in a better prediction of the tunnel behaviour over time. This MASW scan will be repeated at the Heinenoordtunnel and combined with deformation measurements obtained using optical distributed strain measurements (Broere, 2018).

This subcommittee will, just like the subcommittee on joints, carry out the supervision and interpretation of applied research projects (first the Kiltunnel and later the Eerste Heinenoordtunnel and other tunnels) and make the results generally applicable. The results of this subcommittee will be translated via the steering committee into PhD research at universities, applied in practical projects throughout the tunnel programme, and published as knowledge products in the COB knowledge bank.

3. Degradation of materials and components in tunnels

The problem of the degradation of concrete and the corrosion of reinforcement is a major issue in older tunnels. This can lead to reduced strength and leakages in tunnels, as well as damage to the integrity of the structure. There are currently also questions regarding escape doors, fire-resistant materials and fire compartmentalisation in, for example, the Piet Heintunnel.

The renovation of the Maastunnel offers an opportunity to gather more fundamental knowledge about degradation processes through monitoring. Rotterdam has indicated its willingness to allow the Delft University of Technology and COB to set up and carry out monitoring during the renovation of the second tube and use the results for scientific and practical use. The results of this subcommittee will also be translated via the steering committee into PhD research at universities, applied in practical projects through the tunnel programme, and published as knowledge products in the COB knowledge bank.

2.3 Project 3: Modular renovation

Choosing the right scenario for renovation works and making sure that the chosen scenario is feasible, is a major challenge for tunnel owners and renovation teams on both the client and contractor side. That is why it is wise to combine knowledge, instruments and experiences to develop a framework and blueprint for modular renovation of tunnels. The COB distinguishes three types of modularity:

1. The definition of a 'module'. What do we mean, what kind of modules can we identify? Can we find a shared language?
2. The 'time' module (weeks, weekends, nights): can the module be implemented in terms of work within this time unit?
3. The 'building blocks' module (both for the civil and installation technology): can we install a building block and are we able to open the tunnel again (technically operational and with the permission of the competent authority) within the chosen time frame?
4. The 'construction management' module: are we able to conduct the building logistics activities (supply of materials, numbers of people in the tunnel, etc.)?

2.3.1 Deliverables

Based on expertise from the COB network and input from completed, current and upcoming renovation projects, we want to arrive at an assessment framework and a description of scenarios with all the pros and cons. This project also aims to contribute to the design and implementation of better maintainable tunnel modules in new-build projects so that, after the catch-up effort with the existing tunnels, we can then carry out eighty per cent of the renovation task during regular maintenance.

2.4 Project 4: 'Know your tunnel' handbook

If we do not know exactly how our tunnel was constructed, which materials and components were used and what their current state is, we will not know how to identify the critical parts, nor what we must inspect, where we must adjust the parameters and how to interpret the results. We will not know what and how we need to record during the operating phase either, while this is important for the period thereafter. We see substantial differences in knowledge regarding their tunnels among the various tunnel managers and owners. The correct as-built information is not always available, there is no shared approach for developing a proper scope and taking a 'baseline measurement' at the start of the renovation (what are we going to monitor, inspect and/or look up?).

In collaboration with several tunnel owners with extensive experience of this process (Maastunnel, Kiltunnel, tunnels in The Hague) and with experts in completing new tunnels, the COB wants to produce a 'Know your tunnel' handbook. This can be used as a framework and form the inspiration for all tunnel owners and managers for regular management and maintenance, as well as a source document for preparatory teams for the start of a renovation and the completion file.

2.4.1 Deliverables
1. Creating a table of contents (TOC) for a handbook.
2. Translating the TOC into specific texts, references, and so forth for the tunnel managers involved until they can complete their own 'Know your tunnel' handbook.
3. As a result of 1 and 2, producing a recommendation for other tunnel managers on how to make their own tunnel handbook.

3 DEVELOPMENT LINE 2: DIGITAL TUNNEL TWIN

In recent years digital modelling has become much more important in tunnel projects. Building Information Modelling (BIM) is now common practice in both renovation projects and in new construction projects for tunnels. The development of digital modelling, gaming, VR, IoT, etc.

requires a different approach, not only during the test phase, but especially during the design phase (model-driven design). Each tunnel project uses a combination of these tools and has its own digital strategy to build and renovate faster, smarter, with less disruption, and to open tunnels faster. This development has its own problems as all innovations relate to challenges such as filling gaps in our knowledge, reducing the time between project phases, managing discrepancies in the requirements of the various stakeholders and resolving gaps in the system. To a large extent, the market and clients can solve this individually, but there are issues for which a joint approach is needed, questions that you want to answer outside projects and challenges that require more time than an individual project allows for. That is why the COB sees it as its task to help, stimulate, exhibit, inform and streamline criticism in this process, especially since all stakeholders agree that the digitisation of every aspect of tunnels has enormous potential.

3.1 What is a digital tunnel twin?

The digital tunnel twin is not a thing, but a very effective tool for designing, building, commissioning, managing, renovating and adapting physical tunnels: better, faster, with fewer breakdowns and added value. The digital tunnel twin is used and developed in all phases of tunnel projects and adds value in every phase and for each stakeholder. The digital tunnel twin includes all developments in the field of virtualisation/visualisation, operating interface, modelling, simulation, testing, gaming and information provision for the tunnel system.

The first steps towards a digital tunnel twin were made by moving from a paper-based to an electronic completion file. Visualisation in 3D-BIM as part of the digital tunnel twin has already become indispensable (and in the future this will be expanded to 6D-BIM). Within the next five years the digital tunnel twin will offer even more functionality: functional and system models will be added to improve virtual testing and allowing emergency services to train and practice in virtual and/or augmented reality.

Figure 2.   Digital tunnel twin is an 3D-BIM model with more functionality.

3.2 COB's role

The COB's role is to boost, facilitate and present information and not to develop, design or set frameworks. It aims to answer questions such as:

1. What is the significance of a digital tunnel twin in the tendering and design phases of projects?
2. How can virtual testing help to limit public disruption?
3. How can a digital twin help with the opening permit, competent authorities and emergency services?
4. How can a digital twin help keep availability as high as possible through smarter maintenance during the operational phase?

5. Which digital twin concepts are used per project and asset life phase?
6. In which phase does each concept of the digital twin originate?
7. What standards, guidelines and best practices are there for each concept of the digital twin?
8. How can a digital twin contribute to education, training and practice?

The COB has initiated four projects to support the development of the digital tunnel twin:

- Project 1: Opening tunnels without problems using the digital tunnel twin.
- Project 2: Opening tunnels faster with the help of virtual testing.
- Project 3: Virtual education, training and practice.
- Project 4: From paper-based to fully digital.

3.3   *Project 1: Opening tunnels without problems using the digital tunnel twin*

We want to increase the probability of problem-free opening of both new construction and renovation projects by disseminating knowledge about and making the best use of sufficiently mature digital and virtual tools: the digital tunnel twin. This project also wants to contribute to organising the dialogue about and support for (elements of) the digital tunnel twin with important stakeholders such as competent authorities, emergency services and managers.

3.3.1 *Explanation*

Opening a tunnel without problems is mainly about finding support among stakeholders such as competent authorities, being able to coordinate with all stakeholders and testing and checking (validating and verifying) long before the tunnel is operational. It is key that stakeholders must be able to trust that the digital twin corresponds to the physical tunnel to be build.

A digital tunnel twin makes it possible to test the requirements, design and operational scenarios before starting the construction or renovation of the tunnel. It is not envisioned to demonstrate (verify) the performance requirements, such as required air and water flows, overpressure, lighting levels, keeping the safe areas free of smoke, and so on. Nor can it be used to test whether all systems are installed properly. Physical testing will therefore always play a role; virtual and physical testing continue to exist side by side.

Actual meaning for a problem-free opening:

- Using a digital tunnel twin, the (operating) processes and system behaviour of tunnels can be made transparent and suitable for coordination in the early stages of a project. Tunnel managers, road traffic controllers and tunnel operators can be involved early in the process, which makes it possible to include their feedback in the final design phase. Competent authorities, safety officers and emergency services are offered insight into (tunnel) processes and can respond to and anticipate them. Feedback and early involvement would ideally lead to acceleration in the acceptance/testing of the operation. It is possible to simulate the system model and compare it with models of other projects or with a reference model. This allows for the testing of the quality and completeness and makes an important contribution to the verification, validation and acceptance even before installation work takes place at the project site.
- The use of a digital tunnel reduces testing at the project location. The on-site work has been reduced to assembly, commissioning and verification of performance requirements. The validation of processes and system behaviour takes place at an earlier stage. The regular alignment with the competent authorities, emergency services, tunnel managers and safety officers can also take place early on.
- Automated testing programmes will identify non-conformal system behaviour and human errors, especially if these programmes are enriched with experiences from previous projects, thus increasing overall quality.
- By adding gaming functionality, it becomes possible to test and tighten scenarios, and to educate, train and practice without the tunnel being physically available (and not just during the opening). It also has a positive effect on incident handling, as emergency services familiarise themselves with the specific characteristics of the tunnel and can practice their responses (in a

virtual environment) before the tunnel is built and throughout the entire lifespan of the tunnel. If there is enough confidence in the simulation, the opening permit may be issued faster.

3.3.2 Manual

As a first deliverable, the COB will produce a manual including materials and support that will help tunnel projects understand:

1. the definition of 'hassle' for the most important stakeholders (the thirteen most important stakeholders are: competent authority and safety region, safety officers, tunnel managers, directive bodies such as LTR and Building Decree, client's internal services, traffic controllers, road managers, technical managers, client's project organisation, contractor's project organisation, parent companies and subcontractors/suppliers);
2. what stakeholders' needs are, in order not to encounter, experience or cause any hassle;
3. the available tools and systems from the digital tunnel twin that can reduce or even prevent this hassle;
4. the digital tools already used at projects;
5. how these or other elements of the digital tunnel twin can improve the process towards problem-free opening (which processes and tools are available, mature and of added value?).

This will also allow for the sharing of knowledge among the six new construction projects.

3.4   Project 2: Opening tunnels faster with the help of virtual testing

We want tunnels to be opened faster by making the best possible use of virtual testing. The basis for this is the 2017 exploratory study 'Shortening tunnel closures through virtual testing', which lists the possibilities and preconditions (COB, 2017b).

3.4.1 Objectives

This project's objectives include:

1. Formulating unambiguous expectations, definitions and a common 'framework' by the experts in the field (both clients and contractors): what exactly do we mean by virtual testing, how far are we, what do we do and when?
2. Formulating the right preconditions by all stakeholders, in particular the competent authorities.
3. Starting and facilitating a dialogue and building trust with important stakeholders such as competent authorities.
4. Supporting practical projects in thinking about, forming a vision on and presenting the possibilities of virtual testing.

We will look at the possibility of advancing virtual testing in a step-by-step approach. Each of the tunnel projects virtually tests a small element and uses this as proof that it can and does work. By involving as many clients as possible, each project will make a (small) contribution on the technical roadmap and this will benefit all parties.

3.4.2 COB's role

3.4.2.1 Frontrunner group

Almost all (construction and renovation) projects use virtual testing to some extent, but there is not yet a consensus on how we understand virtual testing. Do we have the same point of view? What issues do we encounter? When and how will virtual testing be acceptable for authorities? We will set up a frontrunner group with key people from clients and contractors to discuss issues such as the demarcation of virtual testing, what resources are required, how to describe a virtual model and what tools it needs, standardising test results, how to incorporate it in a digital tunnel twin, include it in contracts, and so forth. The deliverable of the group will be a written recommendation for using virtual testing in tunnel projects.

3.4.2.2 *Competent authority group*

Competent authorities must, in addition to their own role as independent evaluators of their tunnels, also come to understand virtual testing and the systems and working methods involved. By jointly organising this process, we can find out what they need to accept virtual testing as real testing. We can also work together to identify and remove any uncertainties. We can do this by including the group of competent authorities in the technical roadmap for the practical projects, where they can observe the process and discuss issues together with us, etc. This way we hope to garner their support. Here the deliverable will be a written recommendation of virtual testing for competent authorities.

3.5   *Project 3: Virtual education, training and practice*

Keeping everyone up-to-date and 'properly trained' for all possible scenarios is a sheer impossible task. That is why we want to encourage the use of digital tools to further the development of virtual education, training and practice based on the experiences, opinions and ideas of stakeholders (such as tunnel managers, operators, emergency services, competent authorities, security officers, etc.) and help them in formulating their vision and making strategic choices. This will result in written recommendations for virtual education, training and practice for tunnel managers and other stakeholders.

3.6   *Project 4: From paper-based to fully digital*

Several tunnel owners and especially tunnel operators tell us that investing in and managing a complex digital tunnel twin does not suit the size and low complexity of their tunnel. Nevertheless, they do realise that digitisation, (big) data, IoT, sustainability objectives, and so on, can also have a very positive effect on their tunnel management and maintenance. What could be the first small but meaningful step for tunnel managers to benefit from the digital tunnel twin? And what preconditions allow tunnel managers to use the digital output produced by builders during the renovation/construction of tunnels?

3.6.1 *Tunnel managers group*

A group of tunnel managers will look at several issues, including predictable maintenance and, BIM-xD. The group will identify work on these and similar issues being done by clients and look at ways of how to gain access to the results of this work and share any knowledge and experiences. This is expected to lead to a report entitled: 'First exploration: From analogue to digital – the tunnel manager's perspective'.

3.6.2 *Involving market parties*

At a later stage we will invite market parties (and other stakeholders) to enter into a dialogue about the preconditions that tunnel managers want to set for the output from renovation/construction. We thus want to build a bridge between the (digital) information that is available from the construction or renovation of a tunnel and the needs of the asset manager and other relevant stakeholders. This is expected to result in a 'Recommendation for using the digital tunnel twin from a management perspective'.

4   CONCLUSION

The COB developed its tunnel programme for The Netherlands to add value and reduce disruptions. It was envisioned that at least three actual tunnel projects had to benefit from the COB's activities and that the resulting practical benefits are translated into precompetitive knowledge for all parties in the Dutch tunnel industry. This will be achieved through the active involvement of all current new construction and renovation projects, major clients and market parties.

Figure 3.   During the COB congress in June 2018 more than 300 network members joined the 'class of 2018' to start learning in the tunnel programme. (Photo by Vincent Basler).

The COB tunnel programme is currently being implemented with the active participation of more than thirty tunnel projects and tunnel owners, who together with almost sixty market parties and various knowledge institutes such as TNO and the Delft University of Technology manage the projects in the programme.

The development of a long-term vision on tunnels has resulted in a joint scope on what are considered important issues to solve. The challenges are too big and too complex for just one party and we therefore need each other. Moreover, the initial results of the tunnel programme show that collaboration pays off and brings about momentum to tackle other issues.

REFERENCES

Bijl, H. 2011. *40 years passion for underground construction*. The Hague: KIVI, Royal Netherlands Society of Engineers.
Broere, W. 2018. *Installation fibre optic monitoring Heinenoordtunnel* (internal report in Dutch). Delft: Delft University of Technology.
COB 2016. *Long-term vision on tunnels in The Netherlands*. Delft: Netherlands Knowledge Centre for Underground Space and Underground Construction.
COB 2017a. *The work programme*. Delft: Netherlands Knowledge Centre for Underground Space and Underground Construction.
COB 2017b. *Shortening tunnel closures through virtual testing* (in Dutch). Delft: Netherlands Knowledge Centre for Underground Space and Underground Construction.

Tunnels and Underground Cities: Engineering and Innovation meet Archaeology,
Architecture and Art, Volume 10: Strategic use of underground
space for resilient cities – Peila, Viggiani & Celestino (Eds)
© 2020 Taylor & Francis Group, London, ISBN 978-0-367-46878-1

Greater Beirut Water Supply Augmentation Project

A.E. El Abed & E.N. Najem
Cooperativa Muratori e Cementisti di Ravenna, Ravenna, Italy

ABSTRACT: Greater Beirut area has been suffering from shortages in potable water for the last forty years with absence of environmental and water management strategies. The Greater Beirut Water Supply Project aims to overcome this shortage by conveying potable water from a hydroelectric power station near the village of Joun, 30 km south of Beirut, to Hadath reservoirs in the Greater Beirut area, in a sustainable way, by using the TBM excavation technology. Three tunnels of 23.26 km total length will be excavated using two Open Gripper TBMs of d=3.5m with an inverted syphon and a 10 km of twin pipeline. The mucking material from tunnel excavation will be reused as a backfilling for the pipeline. The project: phase 1 (GBWSP) and phase 2 (Bisri Dam Construction), whenever finish, will provide an average volume of 500,000 cum/d of potable water for approximately 1.9 million Lebanese living in Greater Beirut area.

1 GREATER BEIRUT WATER SUPPLY AUGMENTATION PROJECT

1.1 *General Introduction*

The Republic of Lebanon represented by the Ministry of Water and Energy and The Council for Development and Reconstruction (CDR) have lunched The Greater Beirut Water Supply Project- Tunnel and Transfer Line Contract (GBWSP) in 2015 which is financed by The World Bank with a value of approximately 200 million dollars. The joint venture Consultant group Dar Al-Handasah (Shair and Partners) s.a.l. and D2 Consult International GmbH was awarded for the project and Cooperativa Muratori & Cementisti – CMC di Ravenna Società Cooperativa was awarded to be the main Contractor. The project commencement date was 20 November 2015 for 1400 days duration. CMC GBWSP Project Manager and CMC Middle East Area Manager Mr. Paolo Mauri, GBWSP Deputy Project Manager Eng. Amer Elabed and the successful international and local CMC team, on-site and in the offices, are executing GBWSP under the supervision of the Consultant DARD2 and the CDR (the client). GBWSP, phase 1 of the Water Augmentation Project of Beirut, purpose is to convey potable water from Awali River, Bisri village, south of Beirut, to Hadath Reservoirs located in Greater Beirut area. The project consists of the execution of 24 km of tunnels (divided in 3 drives), two inverted shafts, 1 river crossing and 10 km of pipeline. The water will be treated by a Water Treatment Plant (will be executed by others) in Ouardaniyeh site before reaching the reservoirs in Hadath. Tunnel excavation method is considered one of the most sustainable way to cross mountainous area and so it is highly recommended to execute Greater Beirut Water Supply Project. The tunnel will have a final inside diameter (after the permanent lining) of 2.8 m. CMC di Ravenna tunneling team has been working three shifts: day, night and maintenance shifts in order to efficiently perform the job.

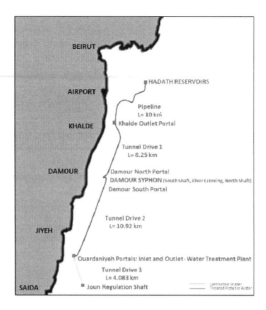

Figure 1.   GBWSP General Map.

1.2   *Tunnel Boring Machine TBM*

In order to perform the excavation of the 23 km tunnels of the Greater Beirut Water Supply Project, two specific Tunneling Boring Machines (TBM) were designed and manufactured by China Railway Engineering Equipment Group CO, L.t.d (CREG) - WIRTH, CREG TBM Germany GmbH. The design of the TBMs adopts double X system of gripper that with the contact on the rock will allow better machine movement and excavation. The characteristics of the two TBMs are:

Table 1.   Tunnel Boring Machines Characteristics.

TBM type	Open Gripper TBM (Kelly Type)
TBM numbers	CREG 237/CREG 238
TBM names	*Lèal Yasmina*
Year of manufacture	2015
Diameter	3.53 m
Number of cutters	26
TBM length	240 m
TBM conveyor length	19 m
Backup conveyor length	130 m
Transformer capacity	2600 KVA

Figure 2.   Tunnel Boring Machine.

1.3 GBWSP Tunnels, Syphon & Transfer Line

1.3.1 Drive 1 – From Damour North Portal to Khalde Outlet Portal

The TBM *Lèa* started excavation of the Drive 1, 5.25 km length, on September, 4, 2017 and it is on-going now with about 75% excavation done and it is expected to breakthrough on December 2018. Along the drive 1, we found one big cavity inside the mountain, W=13m, L=16m, H=13m which has been closed with steel ribs and reinforced concrete encasement following appropriate and approved design. Primary supports (wire-mesh, steel ribs, shotcrete, rock bolts) were applied inside the tunnel where there is bedrock conditions.

Figure 3.   TBM Lèa Assembly in Damour North Portal.

Figure 4.   Inside Tunnel Drive 1.

Figure 5.   Cavity works: from discovery to full closing and resuming excavation.

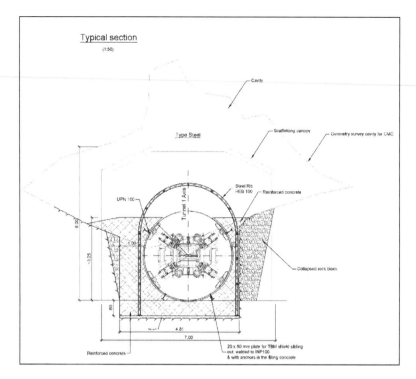

Figure 6. Cross-section of the cavity inside the tunnel and the designed steel ribs and reinforced concrete encasement (the figure is extracted from a full drawing and not to scale in this document, the dimensions are in meters).

1.3.2 *Drive 2 – From Ouardaniyeh Inlet Portal to Damour South Portal*

The TBM *Yasmina* started excavation of Drive 2 with total length of 10.92 km on September 12, 2016 from Damour South Portal and finalized excavation, broke through on March 2, 2018 in Ouardaniyeh Inlet Portal. Primary supports (wire-mesh, steel ribs, shotcrete, rock bolts) were applied inside the tunnel where there is bedrock conditions. A small cavity with stalactite and stalagmite formations was found near the Drive 2 alignment, and a special group of geologists inspected it to study.

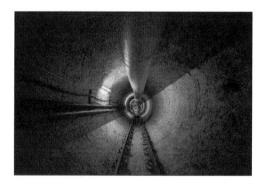

Figure 7. Inside Drive 2.

Figure 8.　Stalactites and Stalagmites inside Cavity parallel to Drive 2.

1.3.3 *Drive3 – From Joun Regulation Shaft to Ouardaniyeh Outlet Portal*

The TBM *Lèa* started excavation of the Drive 3 on June 3, 2016 from Ouardaniyeh Outlet Portal and broke through on May 5, 2017 from Joun Regulation Shaft with a total length of 4.10 km. Primary supports (wire-mesh, steel ribs, shotcrete, rock bolts) were applied inside the tunnel where there is bedrock conditions. The permanent lining works, GRP pipes has started inside Drive 3.

Figure 9.　TBM Lea Breakthrough – Drive 3.　　　　Figure 10.　GRP Lining inside Drive 3.

1.3.4 *Damour Syphon*

Damour Syphon execution is one of the most challenging part in GBWSP. It was designed in the tender stage as 2 vertical shafts of 125m each, which is very risky to execute the raise boring excavation method. For that reason, CMC di Ravenna proposed two inclined shafts (48.18% and 43.62%) to be executed by drill and blasting method using robot drill machines and following the international safety standards of blasting works (proper ventilation, proper lighting, personal protective equipment (PPE), full presence of safety inspectors). The works in these two inclined shafts is ongoing in a safe and successful way. In addition, the syphon includes an under river crossing pipe which was executed successfully under the supervision of the environmental team of the GBWSP to ensure a minimum effect on the river water flow and ecological life. Tunnel supports: shotcrete, steel ribs, wire mesh and rock bolts are applied regularly inside the two shafts of the syphon. A specialized explosives sub-contractor with a certification from Lebanese Government is executing with his specialized team the blasting part of the works.

Figure 11. Inside Inclined Tunnel – Ventilation pipes, proper lighting and wall supports (shotcrete, steel ribs, rock bolts).

Figure 12. Robotdrill machine drilling the holes patterns inside the tunnel syphon.

Figure 13. Removal of mucking materials from the tunnel syphon.

Figure 14.   Pipeline Trench: sand surrounding.

1.3.5 *Transfer Line – From Khalde Outlet Portal to Hadath Reservoirs*

The final section of the GBWSP is the transfer line. Dual ductile iron pipes of 1.4 m diameter each will be placed inside trenches along the public roads, with a total length of 10 km. The execution of the pipeline was done after a full traffic management plan to minimize the interference with the traffic in the affected areas. The pipeline is now about 90% done: excavation, pipe laying, testing and backfilling works. Hadath Reservoirs will be done under a new tendering procedure, (out of GBWSP scope of works).

1.3.6 *Ancillaries works*

GBWSP execution includes many ancillaries' works which has been done by the main contractor CMC. Ancillaries' works include but not limited to:

• Portals permanent and temporary access roads (widening, asphalting, etc.)
• Retaining walls
• Batching plants
• Tunnel wash-out (L=108m) in Wadi Yabes area (executed by drill and blasting method)

Figure 15.   Wadi el Yabes Tunnel Wash-out.

Figure 16. Retaining wall formwork for Joun portal permanent access road.

Figure 17. Base coarse compaction for Joun portal permanent access road.

1.4 *GBWSP highest tunneling excavation records (meters per day)*

Table 2. TBM Lea and Yasmina Highest Excavation Records (meters per day).

Lèa TBM			*Yasmina* TBM.		
Drive	Date	Excavation (m/day)	Drive	Date	Excavation (m/day)
1	17/01/2018	94.67	2	08/01/2018	90.10
1	23/01/2018	85.71	2	18/01/2018	78.13
1	31/01/2018	81.59	2	16/01/2018	74.51
3	06/04/2017	48.23			
3	31/03/2017	48.23			
3	30/04/2017	44.27			

1.5 *Health, Safety and Environmental Criteria of GBWSP*

The Greater Beirut Water Supply Project follows the International Safety and Environmental Standards as per the following:

- Safety Inspectors are always present in all GBWSP working sites.
- HSE training is mandatory for all CMC site and office staff and for all visitors willing to enter the tunnel.
- PPE is mandatory for all site workers. Site entry is not allowed without PPE.
- Safety signs are found outside and inside all sites.
- Proper ventilation is provided inside all tunnels, and it is always inspected.

- Proper lighting is provided inside all tunnels.
- Regular housekeeping for all working sites is mandatory and in a daily basis.
- Following the environmental standards, no dumping sites are found for GBWS because all mucking material resulting from the tunneling excavation is re-used for pipeline backfilling works.
- Replanting of new trees is on-going in all portal sites located in mountainous area.
- GBWSP site offices and compound includes a wastewater treatment system to re-use the water in addition to recycling boxes found inside all offices.
- 24/7 Clinic with a nurse and Ambulance are found in the GBWSP site to respond to any emergency situation.
- The Greater Beirut Water Supply Project has been awarded under the *Vai Sul Sicuro* awards as one of the Safest Sites under CMC di Ravenna Company.

2 CONCLUSION

Greater Beirut Water Supply Project- Tunnel and Transfer line contract is one of the largest construction projects in Lebanon to supply potable water to approximately 1.9 million people living in Greater Beirut area. The project is executed in a very sustainable and environmental friendly techniques, using the most advanced technologies in underground excavation, the TBM machines. Furthermore, more than 200 locals and expats employees are working together as one team in order to efficiently execute the Greater Beirut Water Supply Project.

Tunnels and Underground Cities: Engineering and Innovation meet Archaeology,
Architecture and Art, Volume 10: Strategic use of underground
space for resilient cities – Peila, Viggiani & Celestino (Eds)
© 2020 Taylor & Francis Group, London, ISBN 978-0-367-46878-1

Enabling underground transport construction Guatemala City, Central America

B.R. Fridolin
CEO EMOCON Group, S.A. Ciudad de Guatemala, Guatemala, Guatemala

ABSTRACT: *Current situation, legal and political conditions and development of the first road tunnel project.* Guatemala, Central America (17.57 million inhabitants[1], $10.94 billion budget[2]) demonstrates a pattern of economic, administrative and social concentration in the capital city of Guatemala. The annual net growth of 7% of vehicles in Guatemala[3], is an alarming figure of 5.5 million people circulating on the metropolitan area[4]. This trend contrasts with the growth of 1% of the city's road network[5] and the non-existent alternative of mass transit systems. The result is a daily collapse of vehicles and public traffic.

1 INTRODUCTION

Guatemala infrastructure studies point out to the urgency of underground transport solutions. The State Government of Guatemala as well as the Municipal Government of Guatemala City does not have a long-term progressive plan for underground solutions for urban transport. The private development sector is exploring the possibility of a legal framework for the design of roads and a mass underground transport infrastructure financed, built and operated privately.

National Association of Municipalities of the Republic of Guatemala supports, in 2018, the legal conditions for concessions of municipal underground space. This is a completely new instrument for UG projects improving their viability. The legal approval of municipal governments for projects, concessions of underground roads and mass transit solutions, will be an adequate solution.

The state government does not interfere in the structuring process of the UG project infrastructure. The new Municipal Code law requires competitive criteria such as the low cost for the user, royalty payment, adequate routes, convenience of access and 25 year BOT contract to grant the concession of space for UG projects (*Underground projects*) or road licenses and all permits to private developers.

Guatemala is proposing a new evaluation tool for private underground infrastructure, built and operated, starting with the first project in 2019.

2 WHAT ABOUT GUATEMALA?

Guatemala is the largest economy in Central America with 17.57 million inhabitants and an annual government budget of approximately $10.94 billion.

---

1. DESA UN, 2017.
2. MINFIN-SAT, 2017.
3. DHV Consultants Guatemala, 2018.
4. CIFA USAC, 2005–2006.
5. El Periódico, 2018-Based on Municipality of Guatemala-SAT.

Figure 1.   Guatemala location (EMOCON Group own elaborated, 2017).

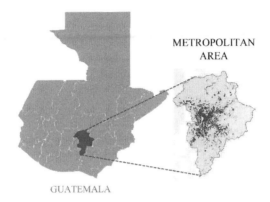

Figure 2.   Guatemala metropolitan area (EMOCON Group own elaborated, 2017).

The pattern of economic, administrative and social concentration Guatemala City, is undeniable. The metropolitan area and this city are territories overcrowded, oversized and extended without any planning, besides having little and inadequate infrastructure. Contrasts with the great territorial of the agrarian area, where the constant of migrants to the Metropolis is 500 p/day, or the United States of America approximately 250 p/day, diminishes all effort and possibility of the state and municipal government for creation of a coherent public infrastructure at regional, metropolitan and city levels.

The social and political agenda of public administration in Guatemala presents a bottleneck situation, financially and politically, and makes it progressively difficult and, in many cases, impossible to allow serious discussions about options and alternatives for viable solutions for urgent urban infrastructure problems. These circumstances strangle the possibilities of innovative and planned design of urban organization for sector of interest.

3   EXISTING CONSTRAINTS

We are detailing existing constraints for coherent action in public administration in general.

3.1   *Politics*

Talking about constraints for solution finding, we first have to look on the agenda of political actors. State Government executes in average only 60 to 70% of state budget. Errant action, with staff technically underdeveloped and opportunistic perspectives, provoke that the institutional agenda is more profoundly oriented towards short term opportunities enforced by peer group actors than long term visionary initiatives.

3.2 *Legal*

The existing legal framework for developing large scale projects is a single privilege of national Congress as the state is the sole agent for providing physical infrastructure in the country and at the same time is the sole owner of underground space according to Civil Law. Since 1965 to 2002, motivated by a constantly underfinanced budget, this privilege was slowly passed to the municipalities as decentralized administrative entities of the State, but never has been expressed within a corresponding legal frame or creating fiscal or financial instruments for the Municipalities to face this responsibility.

As the financial situation of both, state and municipalities, is weak, and the legal frame is insufficient, instruments have to be installed to build visionary infrastructure projects in a pro-active public environment. In 2010 the law for public-private projects was passed, but the institution, ANADIE, which means ironically in Spanish, to nobody, did not allocate one single project until today. The reason is that ANADIE depends on approval of antagonistic interest between Congress, State and Municipal government in an environment eroded by corruptive procedures.

3.3 *Social*

In Guatemala a general social discontent is remarkable during the last 10 years motivated by a general degradation of political culture, increasing inoperative state burocracy and an increasing awareness of corrupt procedures within the state administration. As a consequence we saw a major level of social resistance and two of the four recent presidents in jail. But nonetheless the situation on state level did not improve substantially. Whenever politics gets into infrastructure development, in approval of projects or construction, processes for authorization do not proceed, projects come to their minimum expression and large project do never emerge.

3.4 *Economic*

The financial situation of public funding is widely out of control for several reasons:

- Low capacity of executing public spending in average only approx. 65% of the annual budget.
- High demand for poverty mitigation which is in a strict sense "unproductive" spending.
- Low quality of technical solutions in government projects.
- High prices because of financial costs of projects which is caused by delayed payment.
- High index of generalized corruption.
- Low efficiency of public administration because of 4 years turn-over of administrative staff with each Presidential Election.
- High number of studies for possible projects, but no implementation.
- Extremely low execution of long-term projects especially in infrastructure.

These arguments make the public spending and in general the economic efficiency of public money a marginal exercise.

3.5 *What is the reason for this previous analysis?*

The four core areas are:

- Political segmentation
- Legal inconsistency
- Social disparity
- Economic reason

The four core areas, political segmentation, legal inconsistency, social disparity and economic unreason, in the structuring of the public administration and spending, need to be attended in any kind of alternative project development in the metropolitan and city area especially regarding infrastructure, to find solutions and paths to the planning and development context of traffic and infrastructure. There arises the question, how to enable infrastructure development for mass transportation or, in our case, the development of a traffic city tunnel under the described general conditions?

4 SITUATION OF TRAFFIC AND INFRASTRUCTURE IN THE METROPOLITAN AREA OF GUATEMALA

In the metropolitan area of Guatemala we see a yearly net growth of 7% of vehicles (approx. 70 km/a) and the same percentage regarding the increase of the number of commuting persons (approx. 200 p/day).

This reflects a fast growing traffic in an area with approx. 5.5 million inhabitants today. This tendency contrasts with 1% annual growth of City´s road grid and a virtually non-existing alternative for mass transportation systems besides of a small scale municipal fast track bus line called TRANSMETRO.

Mass passenger transport is now based on a fleet of obsolete buses that do not meet a minimum standard of service and comfort or reliability. Assaults are a daily and sometimes deadly experience for users.

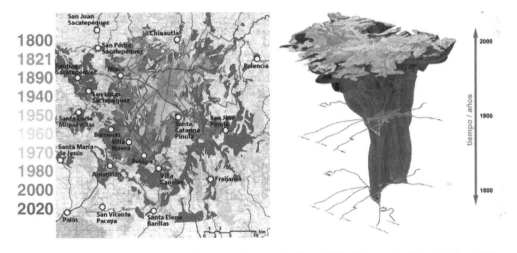

Figure 3.   Population growth metropolitan area, Guatemala City - POT. (Guatemala Municipality, 2010).

Figure 4.   Transmetro. Calzada Raúl Aguilar Batres road (EMOCON Group file, 2017).

Figure 5. Mass public passenger transport service. Trebol – Boulevard Liberación connection (EMOCON Group file, 2017).

Figure 6. Reversible lane. Carretera al Pacífico road - CA9 (EMOCON Group file, 2017).

Figure 7. Road collapse. Carretera a El Salvador road - CA1 (EMOCON Group file, 2017).

Figure 8. Mass traffic Calzada Roosevelt road – CA1 (EMOCON Group file, 2017).

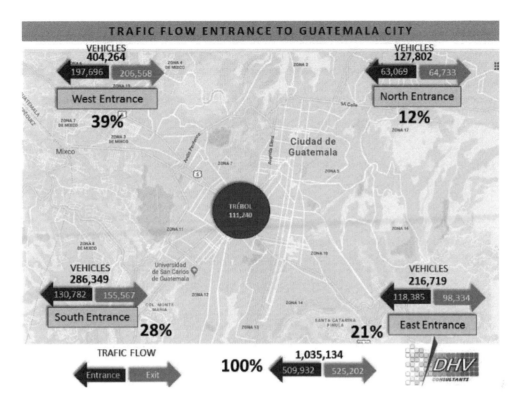

Figure 9. *Diagnóstico de movilidad y entorno Carretera a El Salvador*. (DHV Consultants, 2018).

Car and bus traffic works, in morning and evening peak hours, on the basis of improvised deviating of directional traffic on the opposite lane and installing additional loops for points of high congestion of vehicles.

Result is a daily collapse of vehicle, bus and truck traffic. Previous Studies pointed out the urgency for innovative transportation solutions. Nonetheless Municipal and State governments have no progressive long term plan for underground solutions for inner city transportation, nor for mass neither for road traffic.

The Metropolitan area has four main entrance and exit Corridors for commuting vehicles to and from the City as we can see in that graphic. In the charts we see a compilation of average daily traffic and daily movements of commuters to and from the metropolitan and city area. The average velocity of movement for public buses in the metropolitan area during peak hours is 2.5 km per hour for and 4.5 km per hour for light vehicles according to project investigations.

For the south-east main route which is object of detailed investigation of our development staff, the average velocity is 8.00 km per hour with a medium distance between origin and destiny of 15.00 km. Daily average time budget for commuters in their private vehicle is close to 3.5 hours between origin and destiny, forth and back.

5 TIME AND CIRCUMSTANCES READY FOR PRIVATE INVESTMENT

We, as a private developer for city road infrastructure, think that the time and general conditions are ready to start the first large scale project development and construction for the Guatemala metropolitan area. The criteria for a consistent infrastructure project are:

- Social acceptance for toll collection
- A high proved demand to guarantee the financial feasibility
- Socio-economic income standards
- Technical feasibility
- Financial environment for large scale financing

However constraints are located in the political and legal circumstances. The political environment (government institutions) and Congress (legislation) has a poor reputation as clientelistic and with a sound record of corruptive practices. On the other hand the institution (ANADIE) which is, on basis of law 16-2010, in charge of structuring Private-Public Partnerships (PPP), in 8 years was not successful in allocating one single project, mainly because of a missing technical knowledge, institutional incapacity and the barrier for possible projects of going through Congress for concession approval.

6 STEPS TO A SUCCESSFUL PROJECT DEVELOPMENT

The City of Guatemala road infrastructure on one hand shows a great need for investment, but on the other hand legal, political and financial circumstances are unfavorable for major public investment in road infrastructure.

The private sector, open for opportunities to invest in infrastructure projects, has started to look more closely on this topic. The first point on the agenda was exploring a legal frame for the feasibility of privately designed, financed, built and operated road and mass underground transportation infrastructure.

Several actors have chosen projects which present conditions for private investments. Our company has developed during the past years the "Corredor ESTE – OESTE"- Project which crosses the Metropolitan area from East to West passing through the center district of the City redistributing completely the metropolitan road traffic to and from the Inner City over a total length of 32 km subdivided in four segments with an overall length of 20 km underground development and approx. 12 km of the surface roads.

EMOCON Group is developing at this moment the first segment ("Corredor PONIENTE") which creates a second option to enter or leave the City to and from the South-East direction entering the city from the border to El Salvador. In this area we can observe the City´s most vibrant development with large condominiums (dorm suburbs), a growing service sector to the City (shopping malls and leisure facilities) and a high income average.

Figure 10. Corredor ESTE-OESTE. Metropolitan area east-west (EMOCON Group own elaborated, 2017).

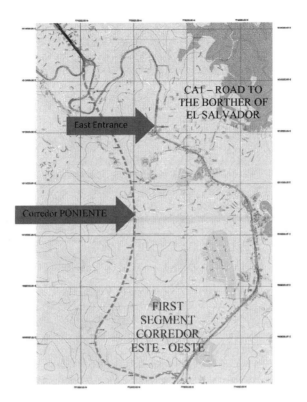

Figure 11. Corredor PONIENTE. First segment Corredor ESTE-OESTE (EMOCON Group file, 2018).

The connecting road to the area, which is approx. 8 Km outside of the City limits and in elevation about 250 meters higher than the City, has only one connecting road which is continuously overloaded with traffic. Approx. 110,000 cars entering or leaving the City daily basis on this road. The presented "Corredor PONIENTE" project, as a private infrastructure project in the Metropolitan area of Guatemala, had to overcome a number of obstacles to achieve pre-feasibility as a viable project. The complete project is of a total length of 7.5 Km, the tunnel segment with a total length of 6.0 Km is planned as a two-tube development with two lanes each, and the surface road segment of approx. 1.5 Km with two double lanes.

The new development bypasses the existing 13.5 Km. South-East road entering the City area which shows continuously congested traffic.

6.1 *Legal viability*

As we pointed out, the established route for private concessions is, up to this moment, to pass through Congress approval which politicizes the process and makes it subject to political favors and uncalculated process costs. The legal solution was to pass the concession process into municipal hands modifying the Municipality functional law or code. With the new legal device, every municipality has the right, within their territory, to allow concessions for infrastructure projects to private investors according to their own interests.

The Developer, the Consultant and Municipalities work together to go through a legislation process of reforming the Municipal Legislation. Fortunately this initiative received the full support of the Association of Municipalities and is about to enter Congress to get approval by March 2019.

This legal instrument is a completely new development device keeping the Central Government institutions and Congress out of the concession process, for enhancing infrastructure projects and especially UG projects on municipal level. Municipal approval of project concessions for underground road and mass transport solutions will be an adequate tool for Guatemala to enhance, besides the large "Corredor ESTE – OESTE" project, a number of UG projects in the future.

6.2 *Technical viability.*

As mentioned, the "Corredor PONIENTE" or the "Corredor ESTE" is a by-pass route to the main road connecting to the South-East region of Guatemala City, and forms also part of the Panamerican Highway connecting to the country of El Salvador. Geological studies show evidence of a satisfactory sedimentary and basalt ground condition. The tunnel center line has an average of 6.5% of decline and over the whole length a minimum of 50 Mts. ground coverage and a maximum of 125 Mts. There is no evidence of abundant ground water. The center line on the surface does not pass under a significant number of buildings. The overall technical conditions for the project show a very positive perspective.

6.3 *Financial viability*

Guatemala as a place for investment has a recognized reputation. Local economy shows a low but steady growth (3.5% over the last 35 years) and there were a number of middle scale projects mostly related to energy production which have created a local expertise in structuring and executing projects.

The Corridor Project as a privately financed and toll based project is currently undergoing a process of evaluation by financial institutions which show interest in funding infrastructure projects in Guatemala. The Net Present Value of the Corridor Project has a sound number and a very promising perspective considering the long term growth rates for road traffic fueled by the high urban development potential for the South-East region of the Metropolitan area of Guatemala. The developing company is at this time negotiating a mezzanine finance to complete the study phase during 2019.

6.4 *Social viability*

The project is located in an urban area so that there are no rural social community issues expected. All neighborhoods in this area see themselves involved in the daily problem of mobility in the area and towards the City.

The three local Municipalities support the project as the concession figure is a BOT model with 25 years of concession time and has included a royalty payment from operation year 10 on.

On this behalf social approval on the local political level is close to 100%. During preparation of the feasibility study during 2019, the project company will publish regularly, in coordination with local newspapers, a number of articles about technical and socio-economic aspects of the traffic situation in the City and a number of publications about the "Corredor PONIENTE" project. As a new element of social mitigation, the Developer has created a

Issuer Credit Rating						
Rating Type	Rating	Rating Date	Regulatory Identifiers	CreditWatch/ Outlook	CreditWatch/ Outlook Date
Local Currency LT	BB Regulatory Disclosures	18-Oct-2017	EE	Stable	18-Oct-2017
Local Currency ST	B Regulatory Disclosures	18-Oct-2001	EE		
Foreign Currency LT	BB- Regulatory Disclosures	18-Oct-2017	EE	Stable	18-Oct-2017
Foreign Currency ST	B Regulatory Disclosures	18-Oct-2001	EE		
Transfer & Convertibility Assessment						
Rating Type	Rating	Rating Date	Regulatory Identifiers	CreditWatch/ Outlook	CreditWatch/ Outlook Date
T&C Assess	BB+ Regulatory Disclosures	18-Oct-2017	..		

Figure 12. Guatemala Standard and Poor´s Rating. (S&P Global Ratings, 2017).

compensation model granting infrastructural benefits to the most affected Municipality of Santa Catarina Pinula.

A special focus will be given to the owners of houses and buildings along the planned alignment of the tunnel. On the other hand the property for the 1.5 Km of surface road will be bought by the project company. As a flexible toll-based project with an alternative route, the commuters have a free choice to take the Corridor route on a toll basis or stay on the existing surface road. Important is that the "Corredor PONIENTE" project gives a very decisive impulse, socially and economically, to the whole Southeast region of the Metropolitan area of Guatemala for future urban development.

7 CONCLUSIONS

The "Corredor ESTE – OESTE" Project as a whole and the "Corredor PONIENTE" Project as the first segment developed, is the first of its kind in the Central American Region, a so-called flag project. The Corredor PONIENTE segment of the larger Corridor Project in Guatemala takes traffic right into the City Center from the South East side of the Metropolitan area. Low technical and financial execution capacity from the Government side, insufficient legal mechanisms and inexistent financial capacity from State side, makes it necessary and possible to create the legal instruments on Municipality level and to develop the project technically by a project developer, receive private financing and cooperate positively with local municipal governments on a royalty basis and to operate successfully the first long range tunnel project in the Central American region.

REFERENCES

Centro de Investigaciones de la Facultad de Arquitectura – CIFA, Universidad de San Carlos de Guatemala (2005–2006). Territorios Urbanos, Programa de Estudios Urbanos y Territoriales. Guatemala City. URL: http://pot.muniguate.com/docts_soporte/14_zonas_especiales.php. (09-01-2019)

DHV Consultants Guatemala (2017). Diagnóstico de Movilidad y Entorno. Guatemala City. P. 145–155

United Nations Department of Economic and Social Affairs – UN DESA (2017). Revisión de 2017 de World Population Prospects. URL: https://countrymeters.info/es/Guatemala (09-01-2019)

Gamarro, U. & Gándara N. (2017). Dictamen de presupuesto 2018. Guatemala City. Prensa Libre. URL: https://www.prensalibre.com/economia/presupuesto-2018-comision-de-finanzas-emite-dicta men-favorable (09-01-2019)

Muñoz, G. (2019). Venta de vehículos nuevos rebasó a la de usados en primer semestre. Guatemala City, El Periódico, P.12.

Tunnels and Underground Cities: Engineering and Innovation meet Archaeology,
Architecture and Art, Volume 10: Strategic use of underground
space for resilient cities – Peila, Viggiani & Celestino (Eds)
© 2020 Taylor & Francis Group, London, ISBN 978-0-367-46878-1

The A2 Maastricht project: A joint action plan

B. Grote & B. Lobbezoo
Avenue2, Maastricht, The Netherlands

L. Prompers & D. Florie
Projectbureau A2 Maastricht, Maastricht, The Netherlands

ABSTRACT: The A2 motorway through the city of Maastricht in The Netherlands proved problematic for several reasons. The international traffic was disturbed by a daily traffic congestion due to six traffic lights. This had a negative impact on the business climate in the city. The A2 also undermined the quality of life, health and safety of its residents. Tunneling proved the best option. The tender was awarded in 2009 to Avenue2, a joint venture of Strukton and Ballast Nedam. They designed and built a double-decker tunnel, 2,3 km long with 4 × 2 lanes. On top, a a pedestrian and cycling route was realized known as the 'Groene Loper' (Green Carpet). The A2 tunnel was opened on 16 December 2016, the exact date specified in the contract.

Challenging in the project was the ground condition with fractured Limestone, layers of Flint and high groundwater table. This lead to the implementation of the Observational Method. The limited working space proved another test as both existing apartment blocks and the A2 motorway were close to the building pit. A different challenge was the economic crisis, the key partners suffered. It even lead to bankruptcy of the tunnel installation company.

1 THE PROJECT

1.1 Background

The A2 motorway through the city of Maastricht in the Netherlands was a thorn in the eye of the city, the surrounding area, and road users for many years until a solution was proposed in 2000. The route through Maastricht proved problematic for several reasons, especially to the international motorway traffic between Maastricht and Southern Europe. The Maastricht motorway was a source of daily traffic congestion due to the six sets of traffic lights. The motorway cut straight through the city, isolating the eastern part of Maastricht from the rest of the city. The A2 motorway greatly undermined the quality of life, health, and traffic safety of nearby residents.

The daily traffic congestion had a negative impact on the business climate in the city and surrounding area. Maastricht and its surrounding area decided decades ago that the 'Berlin wall' (Figure 1) needed to go. Tunneling was by far the best option as it removed most of the traffic from the city and preserved the valuable landscape surrounding Maastricht.

1.2 A Collaborative approach

Due to the common desire to resolve liveability, accessibility, and mobility issues in Maastricht, the governmental stakeholders (the municipality of Maastricht, the Province of Limburg, the national government represented by Rijkswaterstaat [Directorate-General for Public Works and Water Management, the body that maintains the National Infrastructure], and the

Figure 1. Left: The former A2 motorway was long-known as a cause of unsafe traffic situations, poor air quality and noise pollution. Right: The A2 motorway split Maastricht in two like the Berlin Wall.

municipality of Meerssen) opted for a collaborative approach. Together, they commissioned the A2 Maastricht project, contributed according to financial means, shared the risks, and collaborated to create Projectbureau A2 Maastricht and give it a broad mandate to act as the project's commissioning party. An integrated plan was drawn up to commence the assignment which concentrated on achievements instead of requirements (Projectbureau A2Maastricht, 2007). Transforming the section of motorway into a tunnel was a means to multiple ends: diverting the traffic into a two-kilometre-long underground tunnel, developing an attractive residential and recreational area, improving a total of six kilometres of the traffic system surrounding Maastricht by connecting the A2 and A79 motorways at Meerssen, increasing the accessibility of the Beatrixhaven industrial estate in Maastricht, and creating new opportunities for the area surrounding the tunnel (Figure 2). It was an approach that combined improving the infrastructure, integral regional development, and opportunities for the surrounding area. The integral approach was also used in the Madrid Rio, Paris Rive Gauche, and Big Dig (Boston) projects.

1.3 *An Innovative project*

The process was structured in an innovative way as well. The parties that commissioned the project combined legal procedures such as the national government's infrastructure procedure, the municipality's zoning plans, and the selection of contractors from tendering bids in an effort to save time. During the tendering process, they challenged bidders to come up with the best plan by hosting a 'competitively-minded dialogue'. The selection criteria were clearly described, but not set in stone. The structure of the tender to be awarded was called Design & Construct: the contractor both designs and executes the work. The Market actors were therefore given the opportunity to come up with their own solutions. The amount they would receive for the project was pre-determined. Quality was therefore the decisive factor. The winning bid was not only awarded a pre-determined amount of money to design and construct

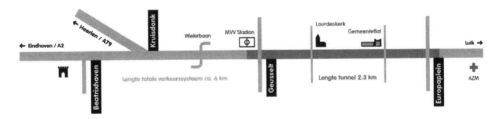

Figure 2. A simplified reproduction of the six kilometers of traffic system for the A2 Maastricht project between the Kruisdonk and Europaplein junctions. The 2.3 kilometer-long tunnel (in dark) is a part of this.

Figure 3.   Left: The press closely followed the project and tender process. Right: During an extra round of consultations, everyone was able to review the plans of the three private companies and respond to them.

the integral regional development and infrastructural works, they were also awarded a property development project including over one thousands houses. The A2 Maastricht project therefore represented a total investment of €1.2 billion euros.

1.4   *Voluntary consultations*

The most unusual thing about the tendering process was that the three competing parties presented their proposals to the general public before a winner was selected. Residents of Maastricht and the surrounding area were able to give feedback on each proposal (Figure 3). Local neighborhood representatives and businesses were also involved in discussions. This lead to Projectbureau A2 Maastricht forwarding approximately 3,500 responses to the bidding parties specifying whether to incorporate, ignore, or consider each piece of feedback. The voluntary consultations with the general public, neighborhoods, and businesses was a bold move by Projectbureau A2 Maastricht that risked endangering the level playing field of the tender. Thankfully they maintained a level playing field being by diligently coordinating the consultation process. This also resulted in the additional benefit of giving the residents of Maastricht and the surrounding area a sense of inclusion from a very early stage of the A2 Maastricht project.

2   THE WINNING PLAN

The innovative tender was awarded to Avenue2, a consortium of the building companies Strukton and Ballast Nedam, in early 2009. Their 'Groene Loper' (Green Carpet) plan was deemed to be the best of the entries. Avenue2 designed a double-decker tunnel that separates traffic in a main route for international traffic and a parallel route for local traffic. This leads to a better and safer traffic flow as the splitting and merging of traffic already happens before and after the tunnel. It also results in more capacity underground (Figure 4). The result was a better traffic system and created more useable land above the tunnel for public space, living environment and thus better real estate.

   The second aspect of Avenue2's winning bid was the 'Groene Loper', a wide pedestrian and cycling route on top of the tunnel. An important condition in achieving a livable area is the connection of the adjacent neighborhoods and the connection between the northern part of Maastricht and the Geul & Maas country estate area, a place of cultural-historical and topographical importance. With a smaller tunnel and the resulting 'Groene Loper' on top it was possible to create an atmosphere of a Park lane instead of a Park (Op de Laak & Smit 2015). This contributes to the sense of being on a connecting element. The 'Groene Loper' thus adds value by increasing the recreational activities available to the Maastricht residents. Due to the width of double-decker tunnel this new Park lane fits in perfectly with the build-up area.

Figure 4. An illustration of the double-decker tunnel and The Groene Loper, from Avenue2's plan (2009). The lower tunnels for international/regional traffic and the upper tunnels for local traffic.

2.1 *Logistical challenge*

The double-decker tunnel also reduced the size of the excavation, meaning that the motorway could be temporarily moved along the building site. During construction it ensured a continuous flow of 50,000 traffic movements per day and at the same time separating it from construction traffic (Figure 5, left) (Bult & Vossebelt, 2011); an important prerequisite from Projectbureau A2 Maastricht. It offered the possibility of building the tunnel in one flow without many phases.

Limiting the working space still proved to be a challenge as both existing apartment blocks and the A2 motorway were very close to the building pit (Figure 5). Leading to an enormous logistic challenge for the construction management. Special logistical managers and traffic controllers were appointed to direct the construction traffic. Also the diversion of cables and pipes of third parties were well in advance incorporated in this logistical challenge. They took the opportunity of the project to invest in new utility networks.

Figure 5. The N2 motorway was temporarily moved to the west for the construction of the tunnel.

Figure 6. Left: A special monitoring system was used to keep an eye on the tunnel construction pit 24/7 during the excavation work. Right: The double-decker tunnel with all the roads connected to it is unlike most projects. That is why a traffic psychologist evaluated the design as perceived and experienced by drivers.

2.2 *Observational method*

Challenging in the project was the ground condition with fractured Limestone, layers of Flint and a high groundwater table. This lead to the implementation of the Observational Method to monitor the progress of the ongoing work during the digging process (Figure 6, left) (van Dalen & Servais & Boone 2015). This precaution was necessary because there was little experience in building a traffic tunnel in the Limestone of the Maastricht area. The Observational Method is a design method which does not deal with the uncertainty of the subsoil by assuming the worst case scenario and applying the full safety factors. Instead the performance is extensively monitored during construction, and for all foreseeable, but uncertain events, a follow up scenario with mitigation measures is present. In this way, the most economical solution in terms of desired reliability level and investment can be achieved. In the case of A2 Maastricht, the Observational Method has proven to be an efficient way to deal with uncertainties regarding the cohesion and permeability of the Limestone.

3 A SUCCESFUL PROJECT

The A2 tunnel in Maastricht opened to the public on 16 December 2016, the exact date specified in Avenue2's contract. The governmental stakeholders did not exceed their budgets. The flow of traffic through Maastricht by way of the temporarily A2 was not measurably worse than before the project started. What made this project to a success, beyond an excellent head start (chapter 1) and excellent plan (chapter 2)?

3.1 *Best for Project*

The project implemented the 'Best for Project' philosophy. Projectbureau A2 Maastricht and contractor Avenue2 maintained a very close working relationship from the very start (Cüsters, 2018). The board of directors and management shared their concerns on a weekly basis. Their basic principle was that the construction works were not allowed to stop, in order to prevent excess expenditure to the project, and inconvenience to the local residents. The mutual view was to keep the budget inside the project instead of spending it on lawyers. On a continuous basis the risks were closely monitored and agreements were made on how to deal with additional work. The Best for Project philosophy is evident in many aspects of the A2 Maastricht project and was implemented from management to construction worker. It also involved an organizational structure with an integrated project team instead of several 'companies' within the project.

The way in which the permits were granted is a standout example complementing the 'Best for Project' philosophy. The A2 Maastricht project required 640 permits in total. All the applications were processed with machine-like efficiency thanks to the clever way of managing the applications. In addition, no notices of objection were submitted against the permit applications.

The road to success was also paved with difficulties such as in 2015, when both building companies that comprise Avenue2 suffered financial setbacks due to the economic crisis in The Netherlands. It even lead to bankruptcy of the tunnel installation company, facing Avenue2 with a huge organizational problem.

3.2 Ready to operate

The project decided to incorporate the stakeholders, who would use the tunnel, well in advance of the actual opening. A Ready to Operate team was created to facilitated and organise it. Learning, training and practising the teams with a virtual build tunnel combined with practice in reality proved a very efficient quick start for the operating phase. Also the participation of emergency workers in the project proved a valuable addition to the builders.

One of the decisions made by the Ready to Operate team was to remotely operate the tunnel in a traffic centre some 80 km away. The A2 Maastricht tunnel is one of the first in The Netherlands that really was operated in such a way from the very first start. This had great operating advantages for all the stakeholders involved.

Another decision was the employment of traffic psychologists to help flesh out the road designs (Figure 6 right). They performed a Road Safety Audit and Human Factors analysis (Kenjić, 2017). Given the complexity of the traffic system, this analysis enabled the identification and timely prevention of the road safety problems/risks. The measures identified were implemented in one of the successive (design) phases of the project. It proved to be a meaningful contribution to a sustainably safe traffic system of A2 Maastricht.

Since the opening of the tunnel in December 2016 no mayor failures to the tunnel system have occurred, it functions all to plan according to the Dutch Road Authority.

3.3 Communications

The collaborative communication strategy was an another excellent example for the A2 Maastricht project's 'Best for Project' philosophy. Projectbureau A2 Maastricht focused heavily on communicating with the surrounding area from the start of the project. During the tender phase this lead to a sense of inclusion from the public (par. 1.4). To intensify the communication with the public during the realization phase, Projectbureau A2 Maastricht and Avenue2 combined their means of communication in early 2012 and started to communicate with one voice. Something very unusual in projects of this scale. Both teams and communication budget were combined. It paved the way for more focused communication which improved the inclusion of the general public in the A2 Maastricht project story in several ways. Stakeholders received weekly updates in video format on the excavation site and how work was progressing (Figure 7 right). Also weekly progress messages and information bulletins were sent out by post to stakeholders and local newspapers. If required the stakeholder managers informed the stakeholders in person, creating a sense of working together on the project.

Each year the construction site was open to the general public for a weekend (Figure 7 left). Numerous presentations were given in the Information Center A2 Maastricht. 120,000 people in total visited the A2 Maastricht project.

The initial fears from the residents themselves regarding such a major building project in the city slowly changed into admiration. By including them in the construction works residents came to accept the project with open arms. It gave them a great sense of pride and awareness of what they as residents could do for their town. In the course of time it became their tunnel and their Green Carpet. An important effect of this extensive external communication was the internal awareness of the importance of stakeholder management. As a result it turned out to be an improvement in the internal communication.

Figure 7.   Left: Many people visited the construction site during the annual open day. Right: Construction workers shared stories about the ongoing work in the weekly information section on regional television.

3.4   *Attention devoted to people*

The A2 Maastricht project encompasses the stories of many people as well. In spite of the scale of the construction works, special attention was paid to initiatives and special requests submitted by the local residents. The Ronald McDonald-house in Maastricht was granted the opportunity to organize a fund-raising dinner (Figure 8). The priest of the church next to the tunnel route introduced Barbara as the patron saint of tunnellers and construction workers. Machinery was turned off during weddings and funerals held at the church. Construction site watchers, often local pensioners, were unhindered as they eagerly followed the progress of the project. In cooperation with the Tourist Office, tours on the building site and through the tunnel were organized. The guides were trained and informed by the project in order to give them the necessary insight information. Terminally ill patients who may not have survived to see the tunnel completed were given a tour of the tunnel at their request.

Figure 8.   A fundraising dinner in the tunnel for two local Ronald McDonald houses.

4 CONCLUSION

4.1 *Lessons learned*

- By defining the achievements of the project in terms of social added value (e.g. a liveable area) instead of stating requirements it resulted in solutions that otherwise would have been made impossible.
- The four governmental parties created a joined project office and gave it a broad mandate. This ensured a strong project management on the client side. For Avenue2 the joined project office proved very beneficial to avoid being driven from pillar to post.
- The project implemented the 'Best for Project' philosophy. Basic principle was that construction works were not allowed to stop. This resulted in the on time opening of the tunnel.
- The tendering process involved a presentation and a call for feedback from the general public, before a winner was selected. Besides a better plan it also resulted in giving the residents of Maastricht a sense of participation right from the start.
- To intensify the communication with the public, Project bureau A2 Maastricht and Avenue2 combined their means of communication and started to communicate with one voice. This lead to a more focused communication that improved the inclusion of the general public in the A2 Maastricht project and the stakeholders in particular; it became their tunnel.

4.2 *Inspiration*

Although the public space has been completed in spring 2018 (Figure 9), the A2 Maastricht project is not completely finished. During the next few years, Avenue2 will build new sustainable homes next to the Groene Loper (one of the locations can be seen in Figure 10 top right). The utmost care will be given to ensuring that the opportunities afforded by the Groene Loper will be pursued.

Figure 9.   A new, green piece of the city is created on top of the tunnel.

Figure 10. The tunnel entrances at Geusselt, on the north side of the city, and a section of the Groene Loper.

Although this real estate project will be finished in 2026, a positive transformation of the area can already be seen. The integral approach is paying off. The Dutch government calculated in May 2018 the social added value in terms of increase in real estate prices (Tijm et al. 2018). For an area of 500m surrounding the tunnel an increase of €220 million could be allocated to the A2 Maastricht project.

The four commissioning parties will continue to work together in Projectbureau A2 Maastricht to maintain the integral approach for the future. Due to both its innovative approach and unique cooperation inspired by the Best-for-Project philosophy, several authorities praised the A2 project as a source of inspiration for other projects in The Netherlands and abroad. For instance when the King of The Netherlands opened the 'Groene Loper' (Figure 11) Margot Weijnen [Professor at Delft University of Technology holding the chair of Process and Energy Systems Engineering; founding and scientific director of Next Generation Infrastructures and also member of the Netherlands Scientific Council for the Dutch Government] worded it as follows in her opening speech: "A2 Maastricht has internationally set a new standard at realizing social added value with an innovative approach aimed at developing opportunities for society, for the economy and for nature in and around the city. An integral approach for mobility en livability leading to a result that truly creates the whole to be greater than the sum of the parts."

The A2 Maastricht project also won the Schreuders price in 2017 for subsurface construction in The Netherlands (Figure 11). According to the Jury: "The project is inventive due to the contractors idea of stacking the tunnel. This created more traffic capacity than the reference design for the same price! The jury praises the project for the courage shown. The construction is technically advanced in a difficult environment where risk management was paramount. This project will be an example for other cities and will put the underground option on the map."

The A2 Maastricht project has produced a result that few could have anticipated in advance including: resolving a major bottleneck in the Dutch motorway network, re-connecting the city of Maastricht, improving the connection to the green area, reaffirming pride in the city and region, improving an entire traffic system, and even affording new opportunities to residents of nearby neighborhoods.

Figure 11.   Left: Opening of 'De Groene Loper' by King Willem Alexander of The Netherlands. Right: Winning of the "Schreuders Price" in 2017 for subsurface construction in The Netherlands.

REFERENCES

Bult, H. & Vossebelt, G. 2011. Verkeersmanagement tijdens tunnelbouw A2. In *Verkeerskunde 3/2011*
Custers, J. 2018. 'Een project is zo goed als de kwaliteit van de samenwerking'. In *Het A2 Maastricht Model*: 73–77. Maastricht.
van Dalen, J.H. & Servais R. & Boone, D.C. 2015. Observational Method, Case A2 Maastricht, proc International Symposium on Geotechnical Safety and Risk (ISGSR), Rotterdam, 13–16 October 2015.
Kenjić, Z. 2017. Assurance of Road Safety in a Project by implementation of RSA and Human Factors Analysis, Case A2 Maastricht, The Netherlands.
Op de Laak, N. & Smit, M.G.J. 2015. De Leefbaarheidstunnel; A2-tunnel Maastricht als integrale opgave. *Stadswerk magazine 02/2015*.
Projectbureau A2Maastricht. 2007. Ambitiedocument A2Maastricht. Maastricht.
Tijm, J. et al. 2018. How large are road traffic externalities in the city? The highway tunnelling in Maastricht, the Netherlands, *CPB Discussion Paper | 379* The Hague: CPB Netherlands Bureau for Economic Policy Analysis.

Tunnels and Underground Cities: Engineering and Innovation meet Archaeology, Architecture and Art, Volume 10: Strategic use of underground space for resilient cities – Peila, Viggiani & Celestino (Eds)
© 2020 Taylor & Francis Group, London, ISBN 978-0-367-46878-1

Tunnel alignment design of "Ligne 15 Est" in East Paris

D. Herman, N. Gilleron, L. Rew & M. Sénéchal
EGIS, Paris, France

ABSTRACT: Egis and its partners are designing the eastern section of Line 15, which is one of the four new metro lines that will help develop the outer Parisian suburbs. This 23 km long underground section includes 11 stations, 21 technical shafts, two turnout structures and a service tunnel leading to the maintenance centre of the line. This article presents the challenges of the track alignment and how the designer has produced the optimum solution. Firstly, it describes the predetermined design constraints: the station locations, the operational requirements (maximum speed, user comfort), etc. It then focuses on the interactions between the tunnelling, its underground environment and the surface environment: the geological context with soft ground layers including risks of voids, sensitive buildings, other high capacity transportation infrastructure and the crossing of the Marne River. These have led to a deep, sinuous alignment with some structures over 30 m below ground level.

1 INTRODUCTION

The Grand Paris Express is currently the largest urban transport project in Europe. Its purpose is to integrate the outer Parisian suburbs to create the new Grand Paris with new multimodal transport solutions, thus improving the attractiveness of these zones.

The Grand Paris Express consists of four additional metro lines and the extension of an existing one with a total length of 200 km, 90% underground, and 68 new stations mostly linked to existing lines. It is an automatic metro system and will provide a train every 2 to 3 minutes to 2 million passengers every day and service 165 000 companies. The new line 15 will circle around Paris whereas lines 16, 17 and 18 will connect it to airports, research clusters and developing neighbourhoods. The existing automatic line L14 will be extended north and south to connect these new lines to the centre of Paris.

The Société du Grand Paris, a public agency set up by the French government in 2010, leads this project by supervising the construction of the new lines, the acquisition of rolling stock, the development within and around the stations and the design and building of a pipeline of optic fibres along the new lines.

The design of line 15 is divided into 3 areas: South, East and West. The Société du Grand Paris selected Egis and its partners (group koruseo) to design the eastern part: "Ligne 15 Est" (L15E – see Figure 1).

The 13 cities – from Saint-Denis to Champigny-sur-Marne – that will benefit from L15E are densely urbanised. The arrival of the new metro will help develop access to jobs, universities and public buildings such as hospitals.

L15E is a dual-track 23 km long underground section that includes 11 stations (108 m long), 21 technical shafts and 2 turnout structures. It also includes a 1.8 km long section leading to the line's maintenance area and depot and two smaller tunnels linked to line 15 South. Nine of the new stations are linked to metro or urban regional express (RER) lines going straight to Paris. The stations Saint-Denis-Pleyel and Champigny-Centre are both already under construction. The "Ligne 15 Est" is to be delivered by 2030.

Figure 1.   Map of the Grand Paris Express and "Ligne 15 Est".

In the next section, we will see how the connection criterion creates constraints on the tunnel alignment design in addition to the usual criteria of speed and comfort. We will then focus on the tunnelling constraints caused by the underground environment – geological context, existing infrastructure – and the surface environment – sensitive buildings, etc. – before analysing the consequences on the design of line 15 East.

2   DESIGN CONSTRAINTS

2.1   *Station localisation*

Each station of L15E will be linked to existing and future transport networks, as shown in Figure 2.

This means that the stations have not been located to optimise the tunnel alignment, but to have the shortest connection time between the adjacent transportation networks, thus creating a new hub in the city. Therefore the "best" station locations could imply increasing the length of the tunnel and consequently the travel time.

The physical interfaces between existing underground transport infrastructure and the new L15E tunnel need to be as short as possible to reduce the risk of adversely affecting existing or future structures. The urban insertion at the surface is the other major factor in the

Figure 2.   Line 15 East connections with other transport networks (Saint-Denis Pleyel is part of L15 North).

positioning of the stations. These lead to the new L15E stations being in general perpendicular to the existing stations. The L15E tunnel is excavated as close as possible under the existing stations (or tunnels) so that the new stations – linked to the existing – have the shallowest platform depth possible.

When linked to surface transportation networks (tramway, RER), the L15E stations are usually placed parallel to the existing stations to ease the traffic flows.

The combination of these design constraints lead to the station orientations being very different one from another. When the stations are also close to each other, it results in a sinuous tunnel alignment between the stations while respecting track alignment design parameters and avoiding sensitive buildings. For example, the shortest distance between Bondy station and Rosny-Bois-Perrier station is 1 293 m, whereas the tunnel is 1 708 m long including 108 m in station (+32% - see Figure 3).

2.2   Shafts and turnout structures

In France, the Tunnel standard of 22 November 2005 concerning safety inside public transportation tunnels requires the construction of an emergency access every 800 m maximum and requires that every point located further than 800 m from a potential smoke source is safe. It means for L15E that there has to be a ventilation shaft every 1 600 m at most.

Every metro station works as an emergency access point and can ventilate 800 m of tunnel either side. However, shafts are necessary between the stations as the stations are always more than 800 m apart. There are two types of shafts: escape (11 m to 16 m diameter) and escape + ventilation (11 m to 20 m diameter). Their diameters vary according to surface constraints and to their depth (shafts deeper than 30 m require a lift). Most of the shafts need at least 3 500 m² for their construction, space that has to be found in a densely urbanised area. The shaft base is at approximately the same depth as the tunnel invert.

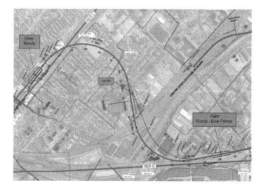

Figure 3.   Tunnel alignment between Bondy station and Rosny-Bois-Perrier station.

The tunnel needs to be close to the shafts so that the underground gallery connecting them is as short as possible. This means that the tunnel sometimes needs to take a detour between stations to be as close as possible to the shafts, whose positions are often defined by the possibilities of surface plot acquisition.

Moreover, the shafts are also used to pump out tunnel infiltration water. Therefore, the tunnel's low points need to be as close as possible to the shafts, which is another constraint on the tunnel alignment design.

Finally, shafts must include the various technical rooms necessary for their operation (electricity, ventilation, etc.). Inserting all of the equipment inside the shafts gives a minimum required design depth of 31.4 m to 36 m for escape + ventilation shafts depending on the equipment needed in the shaft. This also defines the vertical tunnel alignment.

Line 15 East has two turnout structures.

The first one is a 356 m long straight underground structure contiguous to the Rosny-Bois-Perrier station and connects the running tunnel to the service tunnel leading to the maintenance centre and depot. This structure includes a third track used to add or withdraw trains from the running tunnel without affecting the circulation of the other trains, thus a very long structure.

The second turnout is located after Champigny-Centre station and connects Line 15 East and Line 15 South with two single-track tunnels. A third dual-track tunnel enables trains to go back to Saint-Denis instead of going to L15S.

These two turnout structures include several switches and crossings that are additional constraints on the track alignment design.

2.3 *Speed and comfort*

Grand Paris Express is a brand new metro line that aims to be one of the most efficient metro lines in the world. Line 15 trains of will be 108 m long, automatic and will operate at 2 to 3 minute intervals during peak time.

Although the rolling stock wasn't selected at the beginning of the project, its main characteristics such as its height, width or maximum speed were defined to meet the need of the number of users and accommodate the maximum of existing and future rolling stock designs. These features are decisive as an input to choose the tunnel dimensions and the alignment design parameters.

At the same time, the comfort parameters were specified. They define project maximum vertical acceleration, maximum non-compensated transversal acceleration or maximum cant – and its rate of change – etc., so that the user has the best transportation experience. The French standard NF EN 13803 defines these criteria.

Using the rolling stock characteristics and comfort parameters, the designer defines the horizontal alignment parameters – length and radius of curve, length of transition curve (clothoid) – and vertical alignment parameters – gradient, length and radius of curve. The desirable minimum horizontal radius for example depends on the speed, cant and maximum non-compensated transversal acceleration with the formula below:

$$\Gamma_p = \frac{S^2}{R} - \frac{c \times g}{E} \tag{1}$$

where Γ_p = perceived acceleration, S = train speed, R = curve radius, c = cant, g = gravity and E = track gauge.

To reach the maximum design speed (120 km/h for GPE) with the maximum cant (160 mm) and without exceeding the maximum perceived acceleration (0.883 m/s²), the minimum curve radius is 562 m rounded up to 600 m. Using a lower radius results in lowering the train speed to keep the non-compensated transversal acceleration below the limit. The absolute minimum horizontal radius for the running tunnels on this project is 300 m (85 km/h).

A similar method is used to determine the minimum radius of vertical curve.

The minimum slope gradient is 0.5% so that water that infiltrates the tunnel can flow to a low point in the tunnel. The maximum is 4% so that the rolling stock can climb the slope with enough traction.

The horizontal and vertical alignment are interdependent. When combining horizontal curvature with longitudinal gradients, the rolling resistance associated with the trains is likely to result in the train not being able to negotiate a coexisting gradient equal to the maximum specified value. It means that the maximum allowable gradient in a horizontal curve is lower than the maximum allowable gradient in a straight alignment and depends on the value of the curve radius, so that the equivalent slope – Seq' = Slope + (800/R) – does not exceed the desirable maximum gradient.

Furthermore, the vertical curves need to be offset as much as possible from the clothoids on the horizontal alignment to avoid excessive stress in the rails (Figure 4).

Once the main parameters are defined, the dynamic kinematic envelope of the trains is determined for the smallest curve radius used on the project and in a straight alignment. It takes into account the static envelope and its widening through curves, the cant and the minimal clearance between the vehicles. The emergency walkways on either side of the tunnel and the other equipment are then added to obtain the required diameter of the tunnel. Finally, the designer adds the excavation tolerance, the precast concrete segment thickness and the annular void to obtain the excavation diameter of the TBM – 9.80 m for GPE's dual-track tunnels (Figure 5).

With all these parameters, the design of the tunnel can begin. It however needs to be adapted to the underground environment and to the surface environment to control risks during tunnelling.

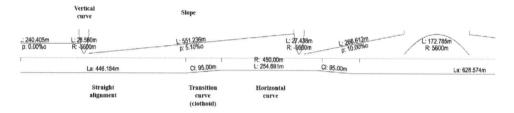

Figure 4.   Extract from the track alignment of the L15E.

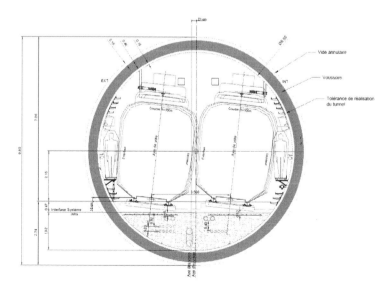

Figure 5.   Sectional view of the L15E tunnel.

3 TUNNELLING CONSTRAINTS

Our design concept was to find the optimum compromise between guaranteeing a minimum height under the surface, to ensure safe boring without affecting sensitive surface buildings, infrastructure and service networks or the environment, and the shallowest possible depths of the stations and shafts.

3.1 *Parisian geology*

Société du Grand Paris has carried out many geotechnical investigations since 2012, providing about 300 borehole results, thus improving the knowledge of the geological context of the ground in the project area. This data helped establish a geological profile centred on the tunnel to calculate among others settlements when tunnelling, thus consolidating the tunnel design.

3.1.1 *Geological context*
The main geological layers encountered in the project area are the following, by order from the surface:

- Soft grounds with low stiffness (< 30 MPa): modern and ancient alluvium, marls with gypsum, soft limestone ("Calcaires de Saint-Ouen" SO);
- Clayey sands ("Sables de Beauchamp" SB) with higher stiffness (around 62 MPa);
- Marly limestone with high deformability variations because of their nature and alteration ("Marnes et Caillasses" MC);
- Coarse limestone that presents high stiffness ("Calcaire grossier" CG).

On L15E the tunnel will mainly be bored through Sables de Beauchamp for the northern half of the route. It will then go through the softer SO between Bondy and Nogent-Le-Perreux stations, before diving under the Marne River and going through harder limestone (MC and CG) until Champigny-Centre. The tunnel will always have at least 1.5 $\emptyset_{tunnel}$ of ground above it ($\approx$ 15 m).

The tunnel will also be mainly bored under a high water load: 20 m on average, but up to 27 m when the tunnel is at its deepest.

3.1.2 *Quarries*
Gypsum quarries can be found in Rosny-Sous-Bois, around 1 km south from the Rosny-Bois-Perrier station. Many sand and limestone quarries were also exploited in Champigny-sur-Marne during the 19th century, at around 15 m depth.

The quarries were backfilled with soft soil and the buildings built on them generally have deep foundations.

As boring under the quarries would mean having to go much deeper, which would require very deep shafts in the area, or could even require backfilling them with concrete, tunnel goes around them as much as possible to avoid sinkholes or voids appearing during the tunnelling.

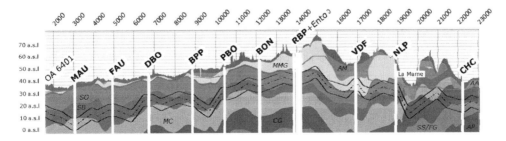

Figure 6.   Overview of the geological longitudinal profile.

3.1.3 *Marne River*

The tunnel passes under the Marne River approximately 1 km south from Nogent-Le-Perreux station. This crossing is a sensitive point where the TBM has to dive with the maximum gradient after leaving the station to be lower than the soft soils (alluvium – AA and AM) and to maintain ground stability. However, a shaft is located about 300 m after the Marne, where the ground is 20 m higher than the river. It means that the tunnel needs to ascend quickly after the river to have the shallowest shaft possible. But the low point of the tunnel also needs to be as close as possible to the previous shaft located before the crossing so that tunnel infiltration water can gravity-flow to the shaft before being pumped out.

All these constraints lead to using a low gradient slope after the low point to avoid the alluvium of the Marne River, which results in shaft 7402P being 37 m deep after the crossing.

The precision of the geotechnical investigations here is very important to maintain the tunnel in the stiff coarse limestone during the crossing.

3.2 *Surrounding structures*

As previously said, L15E will be bored under a densely urbanised area that can be divided into three different types of zone: city centres, suburban areas and industrial areas. City centres tend to have more old and sensitive buildings whereas suburban areas tend to have high residential towers. Each of these factors must be taken into account to control the risks during tunnel boring.

These zones are also traversed by many other transportation infrastructures on the surface and especially underground, and they have also been taken into account during the design.

3.2.1 *Other high capacity transportation infrastructures*

Line 15 East crosses three existing underground metro lines (M5, M7 and M12) and two future extensions (M1 and M11). These crossings happen exclusively near the stations – to create a connection between the lines – where the L15E tunnel needs to be as shallow as possible to reduce the cost of the new stations.

However, L15E will be underneath the other lines for four of the five crossings. Thus the distance between the existing infrastructures and the new tunnel needs to be optimised so that they are as close as possible while controlling the risks.

On this project, this vertical intertube distance varies between 3.0 m (L15E tunnel/M12 tunnel) and 9.4 m (L15E tunnel/M5 station).

The Mairie d'Aubervilliers station is the deepest of the L15E stations (34 m deep) because of the crossing with the M12 whose invert is approximately 23 m deep.

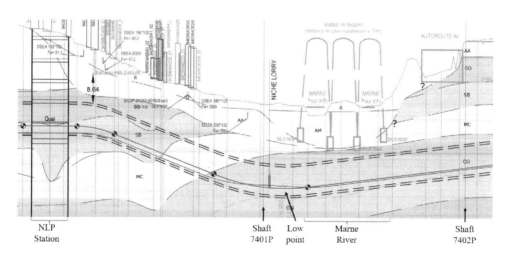

Figure 7. Crossing of the Marne River - Longitudinal profile.

3.2.2 *Sensitive buildings*

Similarly to the geotechnical investigations, Société du Grand Paris initiated surveys on more than 4 000 buildings located near the tunnel to obtain information such as building use (administrative, housing, etc.), its structure, foundations, current state, etc.

Although much data still needs to be acquired, with this information, most buildings are classified by structural sensitivity and global sensitivity. The structural sensitivity represents the building's capacity to withstand external mechanical stresses and depends on the type of structure and its state of conservation. Global sensitivity takes into account the importance of the building for the city: hospitals and schools are considered more sensitive than a sports ground changing room for example. Among the surveyed buildings, for both structural and global sensitivity, more than 82% are considered of "low sensitivity" and only about 1% are considered "very sensitive". The rest are considered "sensitive".

Regarding foundations, 56% of the buildings have superficial foundations, about 2% have deep foundations and 42% are still unknown at this stage.

The surveyed buildings have been integrated in the tunnel alignment design so that the tunnel deviates as much as necessary from sensitive buildings and deep foundations so that the risks when boring and during operation can be managed (see below). This increases the sinuosity of the tunnel alignment.

3.3 *Impacts on boring*

With the geological and hydrogeological data, the designer defines a confining pressure beam for the boring of the TBM throughout the 23 km long route to ensure the stability of the face. Settlements are then calculated, and a more precise confining pressure is selected to control the TBM influence at the surface, taking into account the sensitivity analysis of the surrounding structures.

When approaching singular points such as other underground infrastructure, further calculations are carried out to ensure the safety of the crossing while optimising the tunnel depth when there are significant potential savings.

These calculations sometimes result in an increase of the intertube or building/tunnel distance.

This analysis also shows the minimum necessary cover for launching the TBM from each station and launch shaft. The rails of the shallowest station of L15E, Val-de-Fontenay are about 20 m deep whereas Nogent-Le-Perreux station rails – which has soft grounds above the tunnel – are 27 m deep. The Rosny-sous-Bois launch shaft has been deepened by 3 m after discovering a thick soft layer – 10 m of modern alluvium – above the tunnel.

4 THE RESULTING L15E TUNNEL ALIGNEMENT

The consequence of the various stations' orientations combined with the criteria of track alignment design and the locations of the shafts, explained above, is a very sinuous tunnel alignment, as shown in Figure 8.

The simple measure of the shortest distance between each station, from Stade de France at the north end of the line to Champigny Centre at the south end of the line, is 18.3 km (the green line on Figure 8) whereas the actual length of the tunnel is 20.9 km (the red line on Figure 8, more than 14% longer).

It should be noted that the tunnel alignment between Rosny-Bois-Perrier and Nogent-Le-Perreux stations is almost as short as possible thanks to favourable station orientations. However, on the first half of the project and when arriving at Champigny-Centre, there is a clear discrepancy between the shortest route and the actual design alignment.

The search for the greatest speed while maintaining comfort for the users requires large radii on the horizontal alignment, thus increasing the tunnel length.

On the vertical alignment, the station depths are constrained by either the geological context or existing infrastructure. The shaft depths are mainly related to their belowground equipment

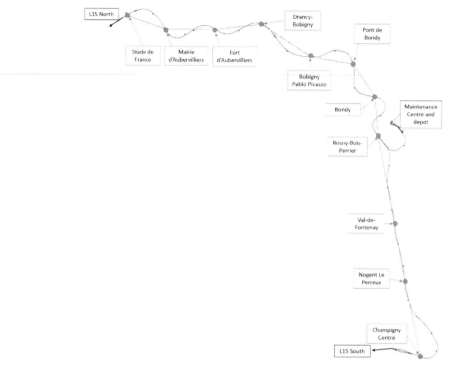

Figure 8.   Comparison between a direct route and the tunnel alignment of the L15E studies.

requirements, but also to the vertical alignment design of the tunnel, that is itself constrained by existing infrastructure, existing foundations and also service networks (sanitation, drinking water, gas). These lead to a deep tunnel with currently one station and nine shafts over 30 m below ground level.

5   CONCLUSION

The tunnel alignment design depends on numerous constraints, either intrinsic: criteria of track alignment design, station and shaft location, or extrinsic: geological context, surrounding structures. These sometimes lead to compromise between the performance of the infrastructure and its costs, but without increasing risks. The following phases of the project will enable to gather additional data on existing buildings, ground conditions and stakeholder requirements and therefore confirm and secure the current tunnel alignment.

REFERENCES

NF EN 13803. 2017. Railway applications – Track – Track alignment design parameters.
Mahdi S., Ningre H., Sénéchal M., Gastebled O. 2019. Grand Paris Express, Line 15 East – Predictive damage analysis combining continuous settlement trough modelling, risk management, automated vulnerability checks and visualization in GIS, ITA-AITES WTC 2019.

Tunnels and Underground Cities: Engineering and Innovation meet Archaeology,
Architecture and Art, Volume 10: Strategic use of underground
space for resilient cities – Peila, Viggiani & Celestino (Eds)
© 2020 Taylor & Francis Group, London, ISBN 978-0-367-46878-1

Operational resilience of traffic tunnels: An example case study

S.S. Khetwal
Ph.D. student, Colorado School of Mines, Golden, USA

S. Pei
Associate Professor, Colorado School of Mines, Golden, USA

M. Gutierrez
J. R. Paden Distinguished Professor, Colorado School of Mines, Golden, USA

ABSTRACT: Functionality losses in tunnels (i.e. partial or full loss of use due to natural and human-induced disruptive events) can greatly undermine the transportation network efficiency. Severe functionality loss related to fire, earthquake or adverse climate will result in significant economic loss thereby affecting the communities socially and economically. Even short-term functionality loss due to minor events such as vehicular breakdown, weather conditions and tunnel repair can also hamper the traffic flow. Tunnel management and operation also affect functionality recovery from the loss, since the tunnel "down time" is highly dependent on the immediate measures taken after an event. Understanding tunnel operational resilience requires a holistic approach considering various scenarios, tunnel type, its location, design, and management methods. The paper discusses the importance of tunnel operation data collection in helping establish tunnel functionality loss models. Using a one-year of operation log data from the Eisenhower-Johnson Mountain Tunnel (EJMT) in Colorado, USA, this study illustrated, using limited data, the approach to statistically determine the operational resilience of EJMT under various events.

1 INTRODUCTION

Functionality losses in tunnels (i.e. partial or full loss of use due to natural and human-induced disruptive events) can greatly undermine the transportation network efficiency. Functionality losses can be either long term or short term, with one caused by major natural or man-made disruptive events like fire, earthquake, adverse climatic conditions, or major renovation; and the other related to routing operational, maintenance or minor traffic breakdowns/accidents. These function loss events can be very costly and have great negative impacts on the public socially and economically. The impact of such events should be quantified when assessing the resilience of the tunnel, with the ultimate objective of keeping transportation services available under adverse circumstances.

Although a unified approach to assess and analyze tunnel resilience is reasonable, majority of past tunnel functionality loss studies analyzed these events on a case-by-case basis. Typically, only significant functionality loss events due to accidents were recorded carefully. There is currently a lack of systematic data collected or analysis done to look into the overall trend for the occurrence and severity of such events. A systematic analysis of tunnel function-loss cases across different levels of severity can answer some of the most critical questions of interest to tunnel owners. For example, in a given tunnel, what is the best and worst-case scenario function loss one can expect, when a certain hazardous condition occurs? Are there certain tunnel type, design, or management methods that are vulnerable to such functional interruption? Is there a statistically significant difference in the recovery time for the same event under different circumstances? This paper statistically determines the operational resilience of

Eisenhower-Johnson Mountain Tunnel (EJMT) under various events based on 1 year of data provided by Colorado Department of Transport (CDOT).

2 LITERATURE REVIEW

Most of the tunnels in the United States were constructed during two eras 0. First after the Great Depression in 1930's and 1940's. Another phase came during the development of Interstate Highway in 1950's and 1960's. Some of these structures have crossed their design service life. They often do not adhere to the latest design codes. Their operation and maintenance periods have increased leading to disruptions in traffic. Regular inspections are necessary to collect data to continue safe operations and prevent structural, geotechnical and functional failures.

An estimation by FHWA data reviles the total length of road tunnels to be 100 miles, approximately 517,000 linear feet—of Interstates, State routes, and local routes in the US (NTIS, 2015). Tunnels accommodate huge volumes of daily traffic, e.g. Lincoln Tunnel between New York and New Jersey carry approx. 120,000 vehicles per day. Amtrak reported an operational loss of nearly $60 million due to the closures of four of its tunnels due to Hurricane Sandy. FHWA estimates that by traveling through the EJMT, the public saved approximately 90.7 million miles of travel per year (NTIS, 2015).

Some major events have led to long-term functional loss. On July 10, 2006; a suspended ceiling panel collapsed on the roadway in I-90 connector tunnel, leading to a fatality and the tunnel closure for 6 months. The traffic was diverted from a longer route, leading to traffic delays and productivity loss. In the Sasago Tunnel in Japan (Kawahara et al., 2014), on December 2, 2012, tunnel ceiling collapsed over a continuous road section of about 140 m, causing nine fatalities. The tunnel was completely closed for 27 days. The commuters had to take a detour, 50 km longer.

There was a major fire in Mont Blanc tunnel on March 24, 1999, causing 39 fatalities and many injured (Fridolf et al., 2013). The cost of repairs and refurbishment with safety upgrades were to the order of $481 million (Barry, 2010). The detour length was around 80 km. In Gotthard Road tunnel on October 24, 2001 in Switzerland (Bettelini et al., 2003) fire broke out, causing 11 fatalities and some injuries. The total repair cost was around $16 million. The tunnel was closed for approximately 2 months. The travel length increased by 30 km crossing the Gotthard Pass, which is susceptible to heavy snowfall and avalanches.

3 RESILIENCE

Resilience is the measure of the ability of a system to resist an unusual disruption and to recover efficiently form the damage state induced by the disruption. For civil structures, resilience of the structure is its ability to function at a certain service level even after the occurrence of an extreme event and to recover to desired functionality as rapidly as possible (Bocchini et al., 2014). Bruneau et al. (2003) gives a conceptual framework defining seismic resilience of communities quantitatively. A measure, Q(t), was defined for the quality of the infrastructure of a community. The seismic resilience is conceptualized as the ability of the system to reduce failure probabilities, reduce consequences from failure and reduce time to recovery. The proposed "resilience triangle" has been used frequently afterwards as the fundamental concept of resilience. Chang and Shinozuka (2004) introduced a probabilistic approach for assessing resilience, measured with loss of performance and length of recovery.

Resilience is a less researched topic in tunneling industry as most of the research is focused in design and construction stages of the project and little study has been done on the operations of tunnels. As the tunnel infrastructure is getting old in the USA and due to climate change, transportation tunnels are under stress from degradation by aging, natural hazards and increasing traffic load. The need for resilient infrastructure has been emphasized in the Presidential Policy Directive (PPD-21, 2013), where a call for, proactive and coordinated efforts, "to strengthen and maintain secure, functioning, and resilient critical infrastructure –

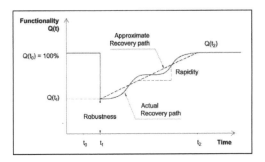

Figure 1. Resilience Triangle.

including assets, networks, and systems – that are vital to public confidence and the Nation's safety, prosperity, and well-being."

There are some studies related to tunnel resilience for specific conditions. Rinaudo et al (2016) defined resilience of tunnel as *"capacity of tunnels to withstand fires with minimum losses and to recuperate a specific tunnel service level as fast as possible."* The paper however, does not quantify resilience in tunnels. Huang & Zhang (2016) proposes a tunnel resilience model, where lining deformation was considered as the metric to quantify tunnel resilience. This model has been developed for the structural resilience of Shanghai metro tunnel lining that is sometimes been subjected to the extreme surcharge loading. Huang et al. (2017) proposed a resilience model for designing repair strategies using real-time monitoring data from wireless network sensors. However, this study is limited to the performance of lining components. In the United States, the Department of Homeland Security has developed Integrated Rapid Visual Screening of Tunnels (IRVS, 2011) for risk assessment and quantification of resilience of tunnels against explosion, fire, and flood. To the authors' knowledge, there has been very few studies that focused on tunnel resilience from end user's perspective, to minimize traffic disruption as well as social and economic losses for commuters and nearby communities.

The objective of improving tunnel resilience is to minimize traffic disruption as well as temporal, social and economic losses. A more resilient tunnel will improve the functional reliability of the highway on which it is located. The resilience of the tunnel will depend on its structural characteristics (like geometry, geological setting, construction technique, support system, aging, etc.), contemporary systems (like Ventilation, Lighting, Fire Suppression, Power Supply), tunnel traffic, operation and maintenance works (including emergency responses) of the tunnel. Substantial research has been done in the field of tunnel maintenance by agencies all over the world. Recommendations on tunnel management and maintenance strategies have been proposed to reduce losses. The previously done research works are mostly qualitative and there is no literature as such, which quantifies the current practices and policies that focus on improving the resilience. Moreover, there is a lack of quantifiable data with the operators or the format of the data is not compatible to quantify the tunnel resilience.

4 OPERATION DATA COLLECTION FOR RESILIENCE ASSESSMENT

Transportation tunnels are complex because they contain many inter-dependent components. The proper functioning of all the components will be required to ensure the full functionality of the tunnel. Some of these components are part of the tunnel infrastructure such as the lining structure and ventilation system, while others can be mobile operational components such as on-site firefighting vehicles. Each tunnel is unique in terms of its components and their protocols in handling disruptive events. The nature and intensity of disruptive events depend on the various factors, like tunnel's geographical location, social behavior of localities, the age of infrastructure, the economic condition of the operator and maintenance pattern. Each event will affect the components of the tunnel in a different manner. Modeling of individual response under hazardous events will be very complicated. Tunnel's recovery after a disturbance is affected by availability of funding, bidding process for repair work, contracting schedule, and other human factors. The assessment of tunnel resilience is very complicated if one plans to use a bottom-up approach, which is to predict the performance of each tunnel component during a disturbance and combine their impact using a fragility-based framework (as it was done for seismic resilience of building structures). Many simplifications must be made, and the results will have a high level of dispersion.

To assess tunnel system resilience a direct approach is proposed in this study. The concept is to systematically collect tunnel functionality loss data during the tunnel operation process using simple metrics such as tunnel lane closure and the time needed for tunnel reopening. The functionality loss of a tunnel is defined as the time-history of tunnel functionality following any disturbance. As it is illustrated in Figure 1 earlier, any functionality loss of a tunnel can be characterized as a series of points in the graph as $[t_1, Q(t_1)]$, $[t_2, Q(t_2)]$,... etc. In this study, the functionality Q, for any traffic tunnel can be quantified as the ratio of traffic capacity available to the public to the maximum traffic capacity available in tunnel design. One simplified way to quantify this metric is:

$$Q = \left(\frac{\# \ of \ open \ lanes}{Total \ \# \ of \ lanes}\right) \times \left(\frac{Reduced \ speed \ limit}{Normal \ speed \ limit}\right) \qquad (1)$$

The advantages of this simple definition are, firstly, Q is solely a state of tunnel operation (i.e. open or close) independent of traffic condition. For example, even when the traffic flow in the tunnel is completely stopped due to traffic jam, if the tunnel is still fully open, its functionality should be 100% (although the efficiency of passengers using this available functionality is very low). Secondly, the simpler the metric is, the easier it is for the tunnel operators to record it accurately every time. The collected data cannot be used directly for resilience prediction, but it can serve as a quantitative measure of existing tunnel resilience against any events that had happened in that tunnel (if such data were collected during that event). Moreover, if one collects a large amount of functionality loss data over the service life of the tunnel, the data set can be combined with other design and operation/management parameters to identify patterns and correlation. This can eventually enable predictive tools supported by large quantity of data.

The challenge with this data-driven approach is that there is currently no uniform data collection strategy or framework. Most of the major tunnels in the U.S. has some level of ad-hoc operation data collection that may or may not contain adequate information for resilience quantification. Hence, the first stage is to establish an ideal data structure for use in everyday tunnel operation management. This framework should be designed to support the assessment of resilience as discussed above, as well as enable further data-driven analysis. A complete tunnel data framework is proposed in this study as shown in Figure 2.

As shown in Figure 2, there are three components in the proposed tunnel data framework. The first being static data, which is defined as the design and construction information which has not changed for a long period unless there is intended changes or upgrades. Static data will include general information, design documents, as built drawings, equipment layout and changes in the layout with time. This information is generally constant over time with minor variations. In cities, the geotechnical setting might change relatively more frequently as the tunnel is close to the surface. Sometimes the condition also changes if an additional bore is constructed close or parallel to the existing tunnel. This data is typically readily available for all U.S. public tunnels through National Tunnel Inventory (SNTI, 2015) and inspection data.

The second component is dynamic data, which is defined as the time-varying condition of the tunnel components and usage conditions. The dynamic data consists of operational and maintenance data of the tunnel components. Collection of this data is a continuous process. Operational data includes traffic volume, status of the equipment, organizational setting, staff and mobile equipment. Maintenance data consist of inspection data on tunnel system and components, which can be used to calculate rate of deterioration, condition states and failure probabilities. The dynamic data can be plotted on a timeline with tags. Some components of

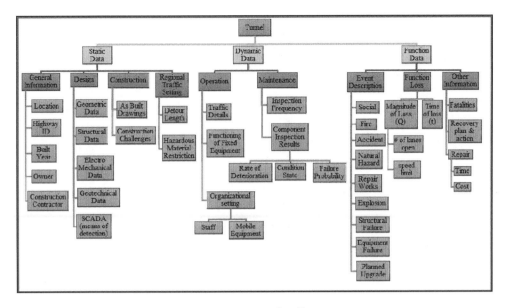

Figure 2. A data collection framework to support tunnel resilience assessment.

the dynamic data are recorded in ad-hoc tunnel management logs kept by individual tunnels, such as traffic volume and operation condition.

The third component is functionality loss data due to an event, which is defined as any information related to reduction and recovery of tunnel functionality (reduced speed, partial closure, full closure). The minimum data collection required is the quantitative measure of available functionality (defined in Equation 1), time stamp (relative to the starting of the event) corresponding to the functionality, and a categorized cause and severity of the external disturbance. The full recovery curve illustrated in Figure 1 can be reconstructed using this minimal information. Additional information that can be included in this category may include fatality, direct financial loss, and cost of restoration. In the analysis later, for EJMT, some of these data were buried partially within ad-hoc logging or traffic management systems not dedicated to record resilience. It is very difficult to extract these data at large scale unless some effort was put towards data collection with this proposed framework in mind.

5 DATA ANALYSIS

An ideal data collection framework is mentioned above, but currently the operators collect data according to their systems developed over time, as per the requirements. At CDOT the tunnel data is collected for EJMT at the tunnel control room. The hourly vehicle count is recorded continuously since the opening of the tunnel, originally by hand and now electronically (still rely on manual inputs). The tunnel does not have an automated system like Supervisory Control and Data Acquisition (SCADA), to record data related to the use of various systems. The operational data is recorded in form of manually generated logs. These logs were initially handwritten on logbooks but are now recorded in excel sheets (still manually). These logs include hourly traffic count, carbon monoxide readings, fan operation status, continuous flow metering and operation activities. These logs are based on visual monitoring via cameras and data collection via sensors. In addition, a recent upgrade in the fire suppression system introduced a new set of sprinklers that are controlled by a separate program. Separate data collection presents a general challenge in older tunnels that went through separate upgrades.

The data related to organization setup of the tunnel is available with CDOT. The tunnel goes through regular inspection checks according to Tunnel Operations, Maintenance,

Table 1. Full Closure events in Eisenhower tunnel from May 2017 to April 2018.

Sr. No.	Direction	Event Type	Event Sub Type	CMV	Fatality	Hazmat	Event Severity	Event Start Date	Event End Date	Full Closure Duration (minutes)
1	East	Planned Event	Road Work	No	No	No	Moderate	Jul 9, 2017	Jul 13, 2017	607
2	East	Incident	Accident	No	No	No	Severe	Apr 21, 2018	Apr 21, 2018	22
3	East	Incident	Mechanical	No	No	No	Severe	Aug 5, 2017	Aug 5, 2017	14
4	East	Incident	Safety Closure	Yes	No	No	Severe	Aug 25, 2017	Aug 25, 2017	73
5	East	Incident	Safety Closure	Yes	No	No	Severe	Dec 25, 2017	Dec 25, 2017	68
6	East	Incident	Safety Closure	No	No	No	Severe	Dec 26, 2017	Dec 26, 2017	14
7	East	Incident	Safety Closure	No	No	No	Severe	Mar 10, 2018	Mar 10, 2018	56
8	East	Incident	Safety Closure	No	No	No	Severe	Apr 24, 2018	Apr 24, 2018	143
9	West	Incident	Accident	No	No	No	Severe	Nov 21, 2017	Nov 21, 2017	14
10	West	Incident	Accident	Yes	No	No	Severe	May 8, 2017	May 8, 2017	9
11	West	Incident	Mechanical	Yes	No	No	Severe	May 26, 2017	May 26, 2017	16
12	West	Incident	Mechanical	No	No	No	Severe	Apr 28, 2018	Apr 28, 2018	3
13	West	Incident	Mechanical	No	No	No	Severe	Apr 28, 2018	Apr 28, 2018	2
14	West	Incident	Outside Agency	No	No	No	Severe	Dec 17, 2017	Dec 17, 2017	3
15	West	Incident	Safety Closure	No	No	No	Severe	May 14, 2017	May 14, 2017	111
16	West	Incident	Safety Closure	Yes	No	No	Severe	Jul 5, 2017	Jul 5, 2017	14
17	West	Incident	Safety Closure	No	No	No	Severe	Oct 1, 2017	Oct 1, 2017	78
18	West	Incident	Safety Closure	No	No	No	Severe	Nov 23, 2017	Nov 23, 2017	18
19	West	Incident	Safety Closure	No	No	No	Severe	Nov 23, 2017	Nov 23, 2017	40
20	West	Incident	Safety Closure	No	No	No	Severe	Dec 7, 2017	Dec 7, 2017	14
21	West	Incident	Safety Closure	Yes	No	No	Severe	Dec 17, 2017	Dec 17, 2017	129
22	West	Incident	Safety Closure	No	No	No	Severe	Mar 4, 2018	Mar 4, 2018	50
23	West	Incident	Safety Closure	No	No	No	Severe	Mar 28, 2018	Mar 28, 2018	36
24	West	Incident	Safety Closure	No	No	No	Severe	Mar 28, 2018	Mar 28, 2018	6
25	West	Incident	Safety Closure	Yes	No	No	Severe	May 18, 2017	May 18, 2017	51
26	West	Incident	Safety Closure	No	No	No	Severe	May 18, 2017	May 18, 2017	50
27	West	Incident	Safety Closure	No	No	No	Severe	May 18, 2017	May 18, 2017	8
28	West	Incident	Safety Closure	Yes	No	No	Severe	Nov 4, 2017	Nov 4, 2017	83
29	West	Incident	Safety Closure	No	No	No	Severe	Dec 21, 2017	Dec 21, 2017	128
30	West	Incident	Safety Closure	No	No	No	Severe	Dec 23, 2017	Dec 23, 2017	11
31	West	Incident	Safety Closure	No	No	No	Severe	Jan 3, 2018	Jan 3, 2018	44
32	West	Incident	Spun Out/Slide Off	Yes	No	No	Severe	Jan 12, 2018	Jan 12, 2018	12
33	West	Planned Event	Avalanche Control	No	No	No	Moderate	Dec 25, 2017	Dec 25, 2017	46
34	West	Planned Event	Avalanche Control	No	No	No	Moderate	Dec 26, 2017	Dec 26, 2017	39
35	West	Planned Event	Road Work	No	No	No	Moderate	Apr 30, 2018	Apr 30, 2018	19
36	West	Planned Event	Road Work	No	No	No	Severe	Jun 10, 2017	Jun 14, 2017	84

Inspection, and Evaluation (TOMIE) manual. CDOT is developing Colorado Tunnel Inventory & Inspection Manual based on TOMIE manual and SNTI to customize it to state specific requirements. Contractors/consultants are hired to do initial, routine and in-depth inspection. Inspection data for structural, geotechnical and electro-mechanical components is recorded in form of inspection logs and reports. These inspection reports can be used to generate most of the maintenance data. However, this process has not been automated.

CDOT is one of the leading DOTs in the US, to modernize data management, initiating statewide management and information system for current and planned deployment of Intelligent Transportation System (ITS, 2008). Moving ahead in the same track, CDOT has developed Colorado Traffic Management system (CTMS) through which the department monitors and records events. The Eisenhower tunnel is part of the system, hence the events happening in the Eisenhower tunnel are also recorded by the CTMS. This might help in corroborating with the data recorded at EJMT. The data structure in CTMS was designed with a focus on traffic accident reporting and resolution instead of focusing on a particular transportation infrastructure component. Thus, it is not dedicated for tunnel traffic alone, but applies to all roadways management by CDOT. Any event that causes disruption of traffic will first be called in either by the monitoring crew or emergency responder. An open call will be generated in CTMS with detailed information about the time, cause, and severity of the event. As the event gets resolved, the parties involved are supposed to report necessary details (time, action taken, resources, etc.) back to CTMS to complete the event data, which gets saved in the CTMS database. CTMS data is a great source for tunnel functionality loss and recovery data.

The data from CTMS was sorted to get the closure information of EJMT. One-year CTMS data was provided by CDOT starting from May 2017 to April 2018. The information recorded in the data made available included partial, full closure or no lanes closed with slowed traffic condition like during continuous flow metering. The direction of traffic which was affected by the event is recorded. The events are divided into type sub-type and sub-class. Information on commercial vehicle (CMV) or hazmat involved in the incident is also recorded. The severity of the event is defined qualitatively as minimal, moderate and severe. The start and end of an event is recorded and hence the duration is calculated. A list of full closure events is given in Table 1.

Table 2. Closure Types at EJMT.

Closure Type	(In minutes)	West Bound	East Bound
Full Closure	Total time	1118	997
	Longest duration	129	607
Partial Closure	Total time	22102	77848
	Longest duration	1768	64364
No Lanes Closed/Slow Traffic	Total time	4796	6269
	Longest duration	2468	538

Table 3. Event Severity in both tunnels at EJMT.

Event Severity	(In minutes)	(West Bound) (Closures)			(East Bound) (Closures)		
		Full	Partial	None	Full	Partial	None
Severe	Number of events	25	8	4	7	4	24
	Total time	1014	1847	349	390	118	4696
	Longest duration	129	983	174	143	46	480
Moderate	Number of events	3	44	6	1	18	6
	Total time	104	20255	1046	607	77730	996
	Longest duration	46	1768	494	607	64364	287
Minimal	Number of events	-	-	5	-	-	3
	Total time	-	-	3401	-	-	577
	Longest duration	-	-	2468	-	-	538

There are 36 full closure events and 74 partial closure events in both bores. The length of closures in the West Bound tunnel more uniformly distributed than the East bound tunnel (Table 2). Although the West bound has 28 full closure events and East bound has just 8 such events. The longer events are generally planned events. There are 74 partial closure events. Again, West bound has more events, 52, as compared to East bound which has 22 of such events. 84% of the partial closure events are planned events, whereas just 14% of the full closure events are planned. 89% of the full closure events are severe whereas 15% of the partial closure events are severe (Table 3). An event type is defined as a planned event or an incident

Table 4.   Event Type in both tunnels at EJMT.

Event Type	Event Sub Type	(In minutes)	(West Bound) (Closures)			(East Bound) (Closures)		
			Full	Partial	None	Full	Partial	None
Planned Event	Road Work	Number of events	2	44	5	1	18	-
		Total time	103	21203	946	607	77730	-
		Longest duration	84	1768	494	607	64364	-
	Avalanche Control	Number of events	2	-	-	-	-	-
		Total time	85	-	-	-	-	-
		Longest duration	46	-	-	-	-	-
Incident	Safety Closure	Number of events	17	-	-	5	-	-
		Total time	871	-	-	354	-	-
		Longest duration	129	-	-	143	-	-
	Mechanical	Number of events	3	3	1	1	1	2
		Total time	21	75	6	14	27	121
		Longest duration	16	35	6	14	27	96
	Accident	Number of events	2	1	1	1	2	2
		Total time	23	74	64	22	76	28
		Longest duration	14	74	64	22	46	14
	Outside Agency Activity	Number of events	1	-	-	-	-	-
		Total time	3	-	-	-	-	-
		Longest duration	3	-	-	-	-	-
	Spun Out/Slide Off	Number of events	1	2	-	-	-	-
		Total time	12	108	-	-	-	-
		Longest duration	12	98	-	-	-	-
	Debris	Number of events	-	-	1	-	1	1
		Total time	-	-	105	-	15	8
		Longest duration	-	-	105	-	15	8
	Abandoned Vehicle	Number of events	-	1	-	-	-	-
		Total time	-	49	-	-	-	-
		Longest duration	-	49	-	-	-	-
	Emergency Roadwork	Number of events	-	1	-	-	-	1
		Total time	-	593	-	-	-	24
		Longest duration	-	593	-	-	-	24
	Snow Removal Ops	Number of events	-	-	3	-	-	1
		Total time	-	-	590	-	-	538
		Longest duration	-	-	373	-	-	538
	Continuous Flow Metering	Number of events	-	-	-	-	-	24
		Total time	-	-	-	-	-	5060
		Longest duration	-	-	-	-	-	480
	Safety Metering	Number of events	-	-	-	-	-	1
		Total time	-	-	-	-	-	287
		Longest duration	-	-	-	-	-	287
	Heavy Traffic	Number of events	-	-	1	-	-	1
		Total time	-	-	100	-	-	203
		Longest duration	-	-	100	-	-	203
	Environmental	Number of events	-	-	1	-	-	-
		Total time	-	-	174	-	-	-
		Longest duration	-	-	174	-	-	-
Weather Event	Warning	Number of events	-	-	2	-	-	-
		Total time	-	-	2811	-	-	-
		Longest duration	-	-	2468	-	-	-

or a weather event, Table 4 shows the event type distribution. These types have subtypes which could be road work, accident, safety closure, mechanical, snow removal, continuous flow metering, etc. Out of 72 planned events only 2 planned events are severe whereas 83 % of the incidents are severe.

6 CONCLUSION AND FUTURE WORK

Tunnels are an important part of transportation infrastructure that are very costly to construct, maintain, and upgrade. To provide a rational approach for tunnel performance, a simple functionality loss metrics is proposed in this study together with the needed data structure to enable the calculation of this metrics. The available data was analyzed as it was provided (in its current state), without calculating the resilience metrics. Although it is difficult to calculate or simulate overall resilience of a given tunnel design, it is hoped that through systematic long-term data collection, the resilience of traffic tunnel can be assessed to justify monetary value of improvements and upgrades in term of resilience improvement. The objective of this study is to suggest a way to collect data to quantify the impact of funding decisions on tunnel infrastructure. While feasibility studies and cost-benefit studies can provide a projected outcome of an investment decision, the data collected following the recommended structure will be able to provide more realistic and quantifiable measures of success.

Observations on current data collection practice of tunnel operation of EJMT in Colorado:

i. The events are logged incomprehensively, as distinct observers log the event differently, thus, making it difficult to extract resilience data. Hence, in this paper the logs were not analyzed. However, the data recorded in CTMS is more organized and has clear descriptors.
ii. The proposed resilience metric is simple to quantify functionality and the severity of events. The metric is defined such that it will be sensitive to upgrades over a long period of time, even if these upgrades have a non-integrated data generation system.
iii. For further analysis, the details of events from the CTMS software are needed to the analyzed and compared to the events recorded in data logs at tunnel level.
iv. The CTMS system used by CDOT has very good potential to become an automated and integrated tool for gathering tunnel resilience data. Although it is event-based rather than infrastructure-based, thereby making it hard to automatically assess resilience of a tunnel. CTMS software can also be upgraded according to the proposed framework.

A tangible metric for tunnel resilience that can be validated by realistic data is of great value to effective management of the tunnels. This study provided a simple tunnel-focused data structure that can be referenced to reduce the fragmentation of data. Once a large quantity of structured tunnel resilience data is collected, it can be used as a basis for efficiency evaluation, cost-benefit analysis, and additional data mining to identify influential factors for tunnel resilience.

ACKNOWLEDGEMENT

The authors would like to acknowledge the financial support from U.S. Department of Transportation (DOT) through University Transportation Center for Underground Transportation Infrastructure (UTC-UTI) (Grant No. 69A3551747118). The opinions and conclusions presented in this paper is that of the authors and do not represent that of the sponsors.

REFERENCES

Barry, K. (2010). July 16, 1965: Mont Blanc Tunnel Opens. *Wired.*

Bettelini, M., Neuenschwander, H., Henke, A., Gagliardi, M. Steiner, W. 2003. The fire in the St. Gotthard Tunnel of October 24, 2001. *International Symposium on Catastrophic Tunnel Fires (CTF), Borås Sweden, 20-21 November 2003.*

Bocchini, P., Frangopol, D. M., Ummenhofer, T., Zinke, T. 2014. Resilience & Sustainability of Civil Infrastructure: Toward a Unified Approach. *Journal of Infrastructure Systems* 20(2),1–16.

Bruneau, M., Chang, S.E., Eguchi, R.T., Lee, G.C., O'Rourke, T.D., Reinhorn, A.M., Shinozuka, M., Tierney, K., Wallace, W.A., Von Winterfeldt, D. 2003. A Framework to Quantitatively Assess and Enhance the Seismic Resilience of Communities. *Earthquake Spectra* 19(4),733–752.

Chang, S. E., Shinozuka, M. 2004. Measuring improvements in the disaster resilience of communities. *Earthquake Spectra* 20(3),739–755.

Colorado Tunnel Inventory & Inspection Manual. 2018. *Colorado Department of Transportation (CDOT).*

Colorado 2035 Statewide Transportation Plan: Intelligent Transportation System (ITS) Technical Report. 2008. *Colorado Department of Transportation (CDOT).*

Ettouney, M., Hughes, S., Walker, R. F., Letvin, E. 2011. Buildings and Infrastructure protection series: Integrated rapid visual screening of tunnels, *BIPS 03, Department of Homeland Security (DHS).*

Fridolf, K., Nilsson, D., Frantzich, H. 2013. Fire Evacuation in Underground Transportation Systems: *A Review of Accidents and Empirical Research. Fire Technology* 49, 451–475.

Huang, H.W., Zhang, D.M. 2016. Resilience analysis of shield tunnel lining under extreme surcharge: Characterization and field application. *Tunnelling and Underground Space Technology* 51, 301–312.

Huang, H.W., Zhang, D.M., Ayyub, B.M. 2017. An integrated risk sensing system for geo-structural safety. *Journal of Rock Mechanics and Geotechnical Engineering* 9, 226–238.

Kawahara, S., Shirato, M., Kajifusa, N., Kutsukake, T. 2014. Investigation of the tunnel ceiling collapse in the central expressway in Japan. *Transportation Research Board 93rd Annual Meeting*, Transportation Research Board.

National Tunnel Inspection Standards(NTIS), (final rule July 14, 2015) *80 Federal Register 41350* (to be codified at 23 CFR Part 650).

PPD-21 2013. Critical Infrastructure Security and Resilience. *Presidential Policy Directive-21.*

Rinaudo, P., Paya-Zaforteza, I., Calderon, P.A. 2016. Improving Tunnel Resilience against Fires: A Methodology Based on Temperature Monitoring. *Tunneling and Underground Space Technology* 52, 71–84.

Specifications for the National Tunnel Inventory (SNTI). 2015. *Publication No. FHWA-HIF-15-006, U.S. Department of Transportation.*

William, B., Steve, E. 2015. Tunnel Operations, Maintenance, Inspection, and Evaluation (TOMIE) Manual, *Publication No. FHWA-HIF-15-005, U.S. Department of Transportation.*

Tunnels and Underground Cities: Engineering and Innovation meet Archaeology,
Architecture and Art, Volume 10: Strategic use of underground
space for resilient cities – Peila, Viggiani & Celestino (Eds)
© 2020 Taylor & Francis Group, London, ISBN 978-0-367-46878-1

Extending the urban underground utility network capacity –
A long drive large diameter pipe-jacking tunnel

A. Koliji
Stucky Ltd, Renens, Switzerland - Current: BG Consulting Engineers, Lausanne, Switzerland

E. Rascol
Swiss Federal Railways, CFF, Lausanne, Switzerland

S. Crisinel
Stucky Ltd, Renens, Switzerland

ABSTRACT: The Western Swiss area has evidenced an increasing urban growth over the recent years. To meet population's need, the Swiss Federal Railways has planned for a strategic use of underground space in Lausanne, one of the major hubs of the region. A new utility tunnel of about 625 m long and 2.5 m internal diameter is constructed, using pipe-jacking and microtunnelling, to reroute the control cables in a newly designed infrastructure at the main train station. Key challenges include tunneling in a very dense urban area and in a geology with possible presence of swelling rock. Time-dependent numerical analyses with advanced elastoplastic constitutive models are carried out to assess swelling behavior of rock. An extensive monitoring of the jacking and TBM performance allowed to reasonably manage the possible risk situations. A close coordination between all urban management parties and services is a success key of the project.

1 INTRODUCTION

Following the extensive economic development of the Lemanic arc region in western Switzerland, the main railway network belonging to the Swiss Federal Railways (CFF) needs to be adapted accordingly. Several very large projects are planned in a short period of time, with challenging constraints in terms of place, time, railway operation and existing construction zones.

The number of train passengers between Lausanne and Geneva has doubled from 2000 to 2010. According to CFF predictions, this number will double once again between 2010 and 2030, increasing from 50'000 to 100'000 passengers per day.

With the purpose of achieving a sustainable solution for population mobility needs, CFF called for a large-scale project - called Leman 2030 - especially dedicated to effective project management and strong coordination with local public representatives and different service departments within CFF. Within this context, the cantons of Vaud and Geneva, the Federal Office of Transport (FOT) and CFF have a common objective – to double the seating capacity of services between Lausanne and Geneva and provide quarter-hourly regional train services by 2030. The implementation of these objectives depends not only on investment in new rolling stock, but also on infrastructure development. There will be new access and transit structures as well as new commercial services in Lausanne, Geneva and other main stations which are currently saturated at rush hour.

The renovation of Lausanne station, a main hub of the region, requires first the adaptation of its railway control center and utility network which is currently out of date for the new developments. A completely new building must be built to accommodate the new equipment and data control systems ensuring safe travel of passengers and cargo trains through the station. The choice of the new location for this building was limited, given the lack of space in the very dense urban area around the train station. A disused cargo station belonging to the national post office was chosen and acquired by CFF for this purpose. Accordingly, all the new control cables as well as the optic fibers and four high voltage had to be re-routed from this new location at the east to the west zone of the station. In such a dense urban area, the most appropriate solution appeared to extend the urban underground utility network capacity through a new tunnel. In the One of the major advantages of an underground solution is that it has relatively limited impact on the environment at surface level, both during construction and in operation, a key issue for a sustainable urban development (Tengborg 2017). This paper presents the planning process and main challenges during the design and construction of the new utility cable tunnel.

2 PROJECT DESCRIPTION

The whole project, including the new control building and the cable tunnel, was approved by Federal Office of Transports in 2016, allowing CFF to start the construction works by June 2016. Given the limited space of only 2000 m2 for construction site, and despite the time constraints, it was decided to plan and construct the tunnel before the new control building.

The tender and final design of civil works for the underground and concrete structures were carried out by the Joint-Venture Stucky-Gruner from 2013 to 2016. Following the tendering in spring 2016, Implenia Construction Company was awarded the contract for construction which lasted until 2018. During this period, the site supervision and construction design was ensured by the designer JV and only the construction design for the pipe-jacking and micro-tunneling was placed under the responsibility of the contractor.

The tunnel has a total length of about 625 m and an internal diameter of 2.5 m to accommodate the utility network cables and provide access for maintenance at all times by CFF technicians. The depth and alignment of the tunnel was defined based on the geological conditions as well as existing and future underground structures surrounding the main station, including metro lines and an existing deep water collector line (See Figure 1). To avoid conflict with other underground structures and utility networks, it was decided to construct the tunnel

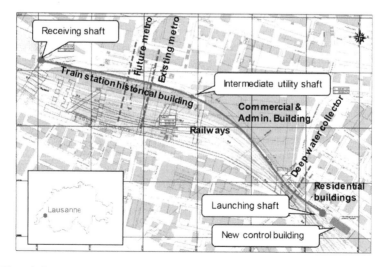

Figure 1. Tunnel alignment and the shafts (Tunnel length about 625 m).

below all existing and future underground structures in the Molassic rock. The final depth of the tunnel ranges from 22 to 27 m below the surface (measured at the axis of the tunnel).

The project includes also two access shafts of 9 and 10 m diameter at the two ends. These shafts will be used first for the construction of the tunnel and then as permanent access structures. They will be equipped with staircases and heavy lifts. There is also an intermediate utility shaft of 1 m diameter which is connected to the tunnel at an underground interception cavern. The cavern is excavated and built after the tunnel completion by enlargement of the corresponding tunnel section.

3 GEOLOGY AND GEOTECHNICS

Geological and geotechnical investigations show a 1 to 6 m thick surface layer of fill material composed of sand, gravel and silt laying on morainic clayey and silty glacial debris. The rock surface is rather regular and is located at a varying depth between 5 m at the east to 12 m on the west side.

The rock on its upper parts is weathered over a zone of 1 to 4 meters thick. Beneath the weathered zone, the existing rock is a rather sound Molasse composed of mainly marlstone with layers of sandstone. The tunnel lies exclusively in the Molasse rock with a minimum rock cover of about 3 times the excavation diameter almost along the whole alignment (See Figure 2). Overall, the geological condition is favorable for the tunnel construction; however, special attention should be paid to possible adverse conditions related to swelling of marl in contact with water.

The ground water table is located at 3 to 5 meters beneath the terrain surface. Hydrogeological investigations indicated varying values of permeabilities, i.e. low permeable parts in marl and potentially high in soil layers. An average water ingress of 100 l/m is expected during the excavation of the shafts.

Geotechnical laboratory tests have been carried out to determine the geomechanical properties of the encountered material including shear strength parameters, stiffness, abrasively and swelling potential. This latter aspect was topic of special attention as the experience from other projects had indicated the presence of black marls with high potential of swelling. The swelling and long-term behavior of the Molasse has been studied by different authors (e.g. Di Murro et al. 2018) and needs to be precisely taken into account for design of underground structures in the region.

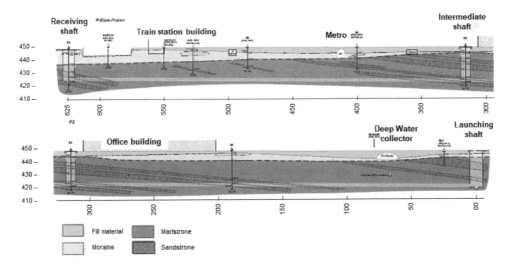

Figure 2. Geological and tunnel longitudinal profile.

Swelling tests of type Huder-Amberg (Huder & Amberg, 1970) were carried out on several samples of marlstone. This test allows determining the two man swelling characteristics, namely swelling strain and the swelling pressure. The highest potential of swelling was observed in a sample of black marl with a swelling deformation of 3.4% after unloading under water soaking.

4 URBAN CONSTRAINTS

4.1 *Urban environment*

The tunnel starts in a residential neighborhood and passes under the railways, a large private office building with three levels of underground constructed on pile foundations, as well as the main train station square and building, which is a classified monument from the early 20th century (see Figure 3). On its way, the tunnel crosses two main road underpasses below multiple train rail tracks bridges. In one of those roads, a deep and large diameter underground sewer water collector is located which is indeed, one of the main constraints for which the tunnel depth had to be adjusted.

In addition, the existing underground metro line is situated 7 m under the surface beneath the railway station. The new tunnel is perpendicular to this sensitive structure. A second metro line, parallel to the existing one at lower level is also planned. The new tunnel is therefore designed to be compatible with the future structure.

The main construction site on the east, including the launching shaft and the tunneling site installation, is in a narrow area between residential buildings and the train railways in operation (Figure 4).

Also on the west side, the shaft is located close to the existing railway control center building. There are, in the basement of this building, all the old mechanical relays controlling the railway system which are sensitive to vibrations. The construction of the shaft and the large open trench induced major disruption for the access of this technical building, which is a key element for the operation of the train station. Many replacement parts are stored in this basement and the access should be maintained 24h over 7 days for different CFF service teams.

Moreover, the heating and cooling systems of the train station are situated in this basement and the pipes leading to the station had to be rerouted during the construction of the trench. An emergency plan was developed with the local head of firefighters, and the organization of this small secondary construction site was strictly restricted to guarantee the fastest possible emergency operation.

4.2 *Coordination with third-party projects*

A close coordination with other ongoing CFF and third-party projects as well as urban management and services was necessary before and during the construction. One of the major

Figure 3. Historical building of Lausanne train station: Left 1925 (courtesy of notrehistoire.ch), right 2018 (courtesy of 24heure.ch).

Figure 4. Launching shaft and main construction site in between railways and residential neighborhood.

parallel projects is the renovation and transformation of Lausanne train station including three new underpasses and a new underground structure below the public square in front of the station (See Figure 3).

Moreover, the western shaft is in the immediate vicinity of the new museum hub regrouping three major museums of the city. The construction of the first new building started at the same time as the construction of the shaft, leading to logistical and political difficulties. However, a proactive coordination and a mutual desire to solve problems allowed to avoid any major impacts for both projects in terms of organization and planning.

5 DESIGN AND CONSTRUCTION ASPECTS

5.1 Pipe-jacking tunnel

The tunnel is constructed using micro TBM and pipe-jacking method. The TBM excavation starts from the launching shaft on the east where the main site installation is planned. Th main advantage of the selected method is minimizing the impacts on the surface and reducing the risk of settlements thanks to immediate installation of lining after excavation.

The pipe-jacking is done in one single drive from the launching to the receiving shaft for the whole length of tunnel (about 625 m) with two curves, the smaller one having a radius of 200 m. At least 5 intermediate jacking stations were foreseen to overcome possible excess friction and enhance the jacking performance.

The concrete pipes have an internal nominal diameter of 2.50 m with a thickness of 250 mm and a nominal length of 3 m. They are constructed with high strength concrete C50/60 and designed for both geotechnical loads as well as jacking forces. Hydraulic joints (Figure 6) are used at the pipe ends to provide a better joint contact for higher jacking forces, and an enhanced performance on small radius curves and long drives. Some selected joints were equipped with pressure transducers and displacement gauges so that the jacking process could be monitored in real-time.

5.2 Shafts

The two access shafts – launching and receiving shafts – are constructed using a mixed solution. On the upper part in soil, a retention system of secant piles is first constructed down to about 2 m into the rock. This retention system ensures the stability of the shaft excavation

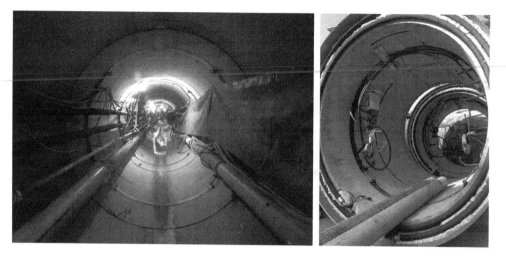

Figure 5.   Pipe-jacking installation: Left - installation inside the tunnel, Right - hydraulic joints.

Figure 6.   Launching shaft: Left – first step excavation within secant piles, Right: TBM installation.

and significantly reduces the risk of settlements and instability in the sensitive environment surrounding the shaft, including the nearby railways in operation (See Figure 4).

The excavation is then started in soil within the pile walls and continued into the rock in steps of about 1to 2 meters. Each step included excavation, shotcreting with wiremesh, circular lattice girders lattices and sealing. Sprayed sealing was used in a sandwich panel between two layers of shotcrete. When water income encountered during excavation, a small pipe was installed in drilled holes to capture water before applying sprayed sealing.

The main jacking station and a pre-tunnel to accommodate the cutter head were the last works in the shaft before launching the TBM.

5.3   *Tunneling in swelling rock*

One of the main challenges of underground works in the Molasse rock of Lausanne region is possible presences of marls with high swelling potential. Design of the tunnel and shafts had to take this issue into account. There are different analytical and numerical approaches to address the swelling problems (Parsapour & Fahimifar, 2016 among others), which should be evaluated and adapted to the project requirements.

If such material encountered during excavation, there would be a direct impact on the design of the shaft structure and the tunnel lining. This impact could be quantified and accounted for in a rather straightforward manner by applying a static asymmetric swelling pressure on the shaft and tunnel lining; and making an appropriate design based on structural analyses.

The other impact, more difficult to quantify, is related to the possible swelling deformation of rock around the TBM during the excavation. An excessive deformation may cause an increased lateral friction, thus, can results in TBM getting stuck with severe adverse consequences on the project planning.

To address this issue, advanced time-dependent numerical finite element analysis has been carried out to evaluate the swelling deformation during excavation and assess the required overcut in the cutter head. Rock excavation and swelling phenomena has been modeled using elastoplastic constitutive model with time and stress dependent swelling behavior. For this purpose, first the model has been calibrated by reproducing the swelling characteristic line obtained from Huder-Amberg laboratory tests (See Figure 7). The material parameters are then used to model the rock around the tunnel based on this calibration.

For modelling the initial-boundary condition problem, several key information and hypotheses had to be considered, as illustrated in Figure 8:

1. Dominant presence of marl on the alignment (12 m)
2. Local presence of black marls with high swelling potential (1 m in the bench)
3. Mobilization of swelling phenomenon around the tunnel over a zone with a width equal to the diameter of excavation (D=3). It should be mentioned that swelling is mobilized only after water reach and therefore, is usually delayed after excavation. The delay depends on both the amount of water and the permeability of the rock. With an estimated permeability of 4.6E-6 m/s, excavation water could reach only 40 cm after one day, and about 3 m after 7.5 days. Accordingly, the following theoretical (but rather safe side) hypothesis are considered for swelling: (a) immediate activation of swelling on 0.4 m during excavation; and (b) delayed excavation on 3 m after 0.5 days (corresponding to almost 50% unloading).

The positive effects of bentonite pressure in the annular gap are not considered. The analyses are carried out for a total duration of 100 days in order to study the swelling behavior during the whole procedure of excavation, driving, grouting and the equilibrium state according to the construction planning.

The result of analyses in terms of time-dependent total displacement (unloading and swelling) on the vault and bench of the tunnel section are presented in Figure 9. These results showed that in case on encountering marlstone at the very beginning of excavation works, the

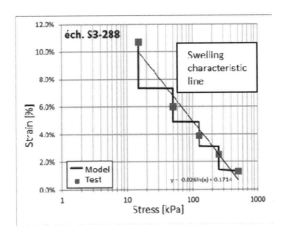

Figure 7.   Huder-Amber swelling model calibration.

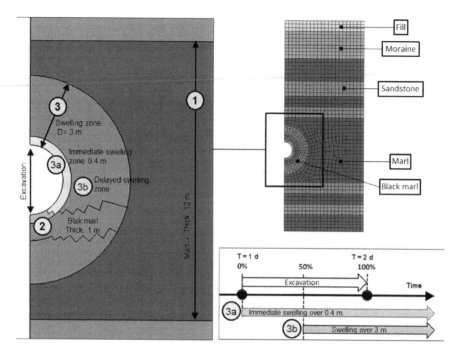

Figure 8.   Tunnel excavation and swelling model hypothesis.

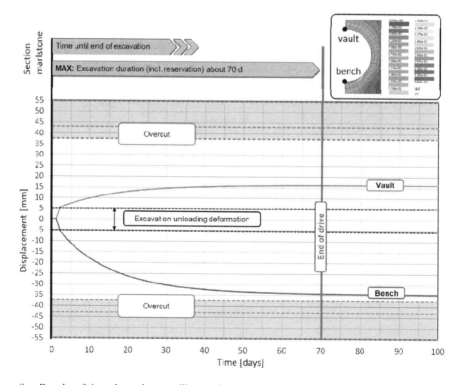

Figure 9.   Results of time-dependent swelling analyses.

expected displacements at the end of the works will be less than 16 mm along the major part of the tunnel. Only on a width of 100 to 120 cm on the bench of the tunnel section, the displacement could reach a maximum value of about 34 mm. Considering that the TBM and the whole string of pipes are floating in bentonite, the obtained theoretical values are rather high limit values. The whole set of analyses allowed to opt for a cutter head diameter with an appropriate overcut to absorb the expected potential swelling displacements during the excavation. The TBM data during excavation revealed no problem regarding the swelling displacements and confirmed the choice of the overcut.

5.4 TBM performance

Extensive monitoring was foreseen to control the performance of the TBM excavation works. Besides the TBM control unit, an additional monitoring system was installed to measure and monitor the TBM parameters including the cutter head position, average speed, torque, RPM, mechanical face pressure, Main jacking force, intermediate jacking force, pipe-string force, average skin friction, and forces in the steering cylinders. Data was processed regularly and daily reports issued. This information is of significant importance not only to monitor the TBM performance but also to identify any risks related to blockage of the TBM or pipe-string due to increased skin friction. In parallel, another data acquisition and monitoring system was installed by the supplier of the hydraulic joints to monitor the pressure and displacements in the selected measurement joints. This information was used to cross-check the pipe-string forces and stress calculation in the pipes.

The tunnel boring took four months, from the 1th of June to the 2nd of October 2017, with two teams working from Monday to Saturday. Detailed data can be found in Table 1.

5.5 Monitoring

Considering the urban environment and the specific requirements of the project, it deemed necessary to implement a systematic instrumentation and monitoring. Measurement of vibrations was continuously carried out at four points among the most sensitive areas, via geophones with continuous measurements over several months.

The most disturbing work was the excavation of some hard rock sections recorded by the monitoring closest to the launching shaft. The two geophones situated in the basement of the large office building, approximately 15 m above the tunnel, never recorded any pick during the tunnel boring. However, the railway traffic made extensive ambient noise. The geophone situated in the room dedicated to mechanical relays also showed small picks during excavation of the hardest rocks.

Four inclinometers were implemented around each access shaft down to a level lower than the bottom slab. They were regularly monitored by an independent engineer during the construction of the shafts. No sign of major displacement was reported in the results.

Table 1. The TBM performance and tunneling data.

Parameter	Value
Total Tunnel Length	617.41 m
Length of precasted concrete tubes	3 m
Number of standard tubes	201
Number of intermediate jacking stations	5
Maximum speed of drilling	Approximately .45 mm/min.
Average advance (including technical stops)	6.5 m/day
Maximum advance	15 m/day
Total duration	103 working days
Machine	Herrenknecht AVN-D
Cutter head diameter	3110 mm

In addition, extensive geodetical survey was carried out all along the project site during the construction. The monitored points included the surroundings of both shafts, the shaft walls, several points along the tunnel alignment at the surface, and the triangulation inside the tunnel, for a total of more than 50 geometrical points. Furthermore, control points in the existing underground metro were taken prior and after passage of the tunnel.

The whole set of monitoring results never showed any excessive displacement on the surface, neither in the existing underground structures, which was consistent with the finite element predictions.

6 CONCLUSION

With a vision of sustainable development, use of underground space in cities become an essential task of resilient urban planning. The utility tunnel project presented in this paper shows a typical case of tunneling with several urban constraints, design and construction challenges. The key aspect for successful construction is the good cooperation of all parties involved including the Owner, Designer and the Contractor.

REFERENCES

Tengborg, P. 1970, *How cities can utilize underground space for sustainable development*, Proceedings of the World Tunnel Congress 2017 – Surface challenges – Underground solutions. Bergen.

Di Murro, V., Li, Z., Soga, K., Scibile, L., 2018, *Long-Term Behavior of CERN Tunnel in the Molasse Region*, proceedings of GeoShanghai 2018 International Conference: Fundamentals of Soil Behaviours pp 671-678.

Huder, J. & Amberg, G. 1970. *Quellung in Mergel, Opalinuston und Anhydrit.* Schweizerische Bauzeitung, 88 (43), 975-980 (in German).

Parsapour, D. & Fahimifar, A., 2016, Semi-analytical solution for time-dependent deformations in swelling rocks around circular tunnels, Geosciences Journal, 20(4), pp 517–528.

Tunnels and Underground Cities: Engineering and Innovation meet Archaeology,
Architecture and Art, Volume 10: Strategic use of underground
space for resilient cities – Peila, Viggiani & Celestino (Eds)
© 2020 Taylor & Francis Group, London, ISBN 978-0-367-46878-1

Improving the safety of urban underground transport areas due to the use of new energy carriers

F. Leismann & F. Baum
Research Association for Tunnels and Transportation Facilities – STUVA e. V., Cologne, Germany

C. Knaust, S. Palis, S. Voigt & F. Sträubig
Federal Institute of Materials Research and Testing – BAM, Berlin, Germany

S. Klüh
FOGTEC fire protection, Cologne, Germany

M. Lakkonen
Institute for applied Fire Safety Research – IFAB, Berlin, Germany

ABSTRACT: To save resources and reduce CO_2 emissions, alternative vehicle drive systems based on New Energy Carriers (NEC), i. e. different types of purely electric, hybrid, gas and bio fuels are increasingly used. These NEC change the risks in transport, especially in enclosed underground areas. Current guidelines do not sufficiently cover the changing risks as only few scientific studies on the impact of NEC on safety issues are available. This gap is addressed by the German SUVEREN project, which investigates phenomena that may occur in connection with the use of NEC in underground urban traffic areas. The research project examines the impact of battery and gas fuels on the safety of enclosed car parks, delivery areas, bus depots and tunnels. The given paper provides insight in this project and its upcoming results leading to a safety concept for appropriate treatment of NEC in underground structures.

1 BACKGROUND AND MOTIVATION

To save resources and reduce CO_2 emissions, alternative drive systems for vehicles have been developed and increasingly used worldwide. Manufacturers offer a wide range of vehicles, which can be powered completely electrically, as hybrids i.e. with a combination of internal combustion motor and electric power, or by gas.

The development of New Energy Carriers (NEC) differs worldwide from country to country depending on various parameters like technical innovation, price development of NEC and conventional energy carriers. Environmental constraints regarding sustainability and pollution, e.g. nitrogen oxide exposure in cities lead to political decisions regarding funding or prohibitions (e. g. Directive 2008/50/EC on ambient air quality and cleaner air for Europe). Accordingly, different scenarios are existing, how the future deployment of NEC will look like [Shell, 2015]. One option limits to an extrapolation of the past situation; the opposite option is a prediction of a massive change due to deterioration of environmental constraints. Independent of the wide scope all scenarios have in common, that NEC will increase significantly. In Germany, due to political consensus e-mobility is expected to prevail, but drive systems based on gas will also be used in a relevant number beside conventional fuels. As a result, NEC as drives and loads for vehicles will increasingly be used also in public and private underground (confined) urban transport infrastructures, so that their impact on safety aspects needs to be considered.

Figure 1. Charging station for e-mobility in an underground parking garage.

The use of new energy sources can cause different hazards compared to conventional fuels, such as battery fires, jet flames from relief valves of pressurized gas containers or the spread of highly flammable gases. Nevertheless, current safety concepts, guidelines or standards for the planning and operation of tunnels and underground spaces (in terms of design basics like heat release rates), rescue concepts and fire-fighting measures) are based on the risks of conventional energy sources, and NEC are not considered yet.

2 PROJECT SUVEREN

2.1 General Approach

To tackle the above-described situation the German Federal Ministry of Education and Research (BMBF) funded the research project SUVEREN in the thematic area "Future Security in Urban Spaces".

Figure 2. Logo of the research project SUVEREN.

Its main goal is to investigate the risks resulting from increased use of new energy carriers (NEC) in vehicles in order to derive suitable safety concepts and to update the safety in underground urban areas. NEC vehicles and the associated risks, e. g. the fire behavior of batteries, pressurized gas containers, and composite materials are investigated. To achieve this goal the project is dealing with the following main tasks:

- Literature studies of past incidents
- Analytical and numerical studies
- Analysis of existing traffic facilities
- Risk analysis of new energy sources
- Development of case studies and reference scenarios for simulation of hazards regarding the use of new energy sources
- Modelling the release of substances and energy from new energy sources
- Modelling of extinguishing processes
- Analysis of hazard scenarios
- Performance of large-scale fire tests to validate the analyses and models

Project partners are the Federal Institute of Materials Research and Testing (BAM), Berlin (coordinator), FOGTEC fire protection, Cologne and STUVA, the Research Association for Tunnels and Transportation Facilities in Germany. Associate partners are DB Station & Service AG, Berlin; City of Munich, Fire Directorate; CETU Centre d'Etudes des Tunnels, Bron (France); and INERIS, Verneuil-en-Halatte (France).

Figure 3.   Hydrogen-Bus.

2.2   *New Energy Carriers (NEC)*

The term „New" in NEC in the given context refers to changing potential consequences and market share that results from substituting conventional energy carrier petrol/diesel. Following this definition, NEC embrace alternative energy carrier (non-fossil), hydrogen gas and natural gas and liquefied petrol (LPG). Further, composite materials are taken into consideration, as these innovative materials also may lead to a change of constraints and resulting risks.

NEC are either used as fuels for internal combustion engines and fuel cells or in the form of batteries as electrical energy. From a safety point of view, new challenges of NEC are chemical processes. Further, in terms of batteries energy densities are rapidly increasing from generation to generation. This leads to new hazard scenarios during operation, parking, recharging or refuelling of NEC.

3   IMPLEMENTATION OF THE PROJECT SUVEREN

3.1   *Risk Analysis*

NEC are very much in the public focus. In case of a burning electric vehicle or an incident with a jet flame from a gas-driven vehicle, the major news report in detail. As a matter of fact, multiple more conventional cars are burning every day but these are mentioned in best case in the local papers. This leads to a distorted public perception that suspects NEC as more dangerous than vehicles with conventional fuels. To clarify this gap of knowledge SUVEREN scientifically researches the level of risk due to NEC compared to conventional fuels.

Risk is defined as the product of the probability of occurrence of an incident and its amount of damage. To quantify the risk due to NEC the first step is to research statistics and critical accidents that took place in the past. Reduced to one sentence the result of the recherché executed in SUVEREN [SUVEREN, 2018] can be summarized, that only few critical accidents have occurred in underground facilities. Single cases are documented, but no overall picture is given. The statistical population is too small to provide a scientifically evaluable base and no reliable statement about the risk of NEC based on the past is possible. Consequently, the identification of relevant scenarios for a reliable prognosis is getting more important.

As second step to calculate the risk the extent of consequences needs to be estimated. NEC as drive or load may cause different risks, e.g. due to battery fires, jetting flames from relief valves of gas pressure vessels, the spread of toxic or highly flammable gases [Gehandler et al., 2017]. These effects need to be determined within SUVEREN in a qualitative and if possible quantitative way. In this context, the environment of the incident plays a major role for the resulting damage. SUVEREN addresses this question by specific case studies, in which NECs in characteristic environments are examined and possible effects are determined.

SUVEREN addresses all reasonable NEC to provide general safety concepts for current challenges. Each energy carrier has specific characteristics. To handle the extensive task within the project the NEC are categorized regarding common characteristics into three main groups.

The first category are gaseous energy sources. Although some of the gases differ significantly regarding properties and application, they react comparably in case of fire. Stored in pressure vessels, during a fire the gas expands to a large extent and blows off via the pressure relief valve. Without ignition, this can generate an explosive gas mixture in the environment. In connection with up to 15-20 m long jet flames [Dutch Safety Board, 2012], depending on pressure and housing, may occur. Apart from different combustion temperatures, storage pressure and physical properties, the overall hazard potential of jet flames is in the same range.

The second category is battery-powered vehicles, where the battery represents a source of danger. Currently, due to best energy storage properties, most car manufacturers use Li-Ion batteries [Zhang et al., 2017]. This type of battery is vulnerable to damages resulting from overheating or mechanical damage. If the battery management system fails or is not able to cope with the damage a thermal runaway of the Li-Ion battery results.

The third category is biofuels used as substitutes for conventional fuels. As biofuels have very similar properties as gasoline or diesel this category has low priority in research.

NEC are present throughout the whole transport sector. To cover the most unfavourable condition SUVEREN refers to underground urban infrastructures, as the confined conditions increase the critical effects (heat, smoke, toxic gases). As underground infrastructures have well defined but very limited escape routes, in the event of fire and smoke – following the concept of self-rescue – users must be able to reach safe areas within a reasonable time. These more critical conditions for users, rescue forces and buildings lead to higher importance of safety measures than above ground. For battery-powered vehicles, relevant factors can be, besides heavy smoke also gas formation, which may occur if Lithium-Ion batteries are mechanically damaged or overheated. The release of different toxic gases, depending on the chemical composition of the batteries, and the associated risks with regard to chemical reactions (hydrogen fluoride) in combination with used extinguishing agents is hard to oversee. Further, a problem for firefighting is the battery housing, which makes access for extinguishing agents very difficult or even impossible.

In case of gas-powered vehicles, new threats result from the emission of various gases with different densities. For gases lighter than air, this can lead to gas accumulation in the ceiling area of underground traffic systems. For gases heavier than air, this can lead to formation of a lake in low-lying areas or to further distribution of gases through drains and sewage systems. In both cases, the question in terms of gas concentration and ignitability of the resulting mixtures is relevant. Regarding the possibility of detection, the question arises which gases in what concentration are expected. Questions regarding the availability of sensors, required sensitivity and the optimum location for the measurement equipment are also addressed within the project.

Together with operators and emergency services, SUVEREN is developing NEC-specific fire scenarios which are included within case studies. The goal is to map the new hazards caused by NEC in a realistic way for five types of underground traffic facilities (case studies). From the defined bunch of scenarios, the ones with the highest reasonable risk (probability x damage) are identified to be analyzed in detail for each underground facility.

For the detailed investigations, the following building types were identified as relevant, as they are already frequently found in the urban areas today or in the near future. Further different protection goals (self-rescue, assisted rescue and asset protection), as well as various fire risks and fire phenomena, are covered by these facilities. The project focuses on:

Underground car parks

Underground car parks are considered to be relevant since they are to be found in all (German) cities and thus the total number is high. Under normal conditions, the number of

Figure 4.   Underground car park.

people inside underground garages or enclosed parking garages is small, which eases evacuation in case of emergency. Due to the individual traffic, underground garages are used by all drive types and energy sources. In public places, with a lot of public traffic, underground car parks may be heavily frequented. The high number of vehicles parked close to each other also causes a high fire load and risk of fire spreading. In the New Year's Eve of 2017 for example, more than 1,400 cars in a Liverpool parking garage burned out completely. In case of fires for the fire brigade, the attack path is particularly difficult due to heavy smoke and fast heat development.

Charging stations for electric vehicles are already installed in many underground garages and big car parks. During charging, the battery is heavily used, which increases the risk of fire significantly.

Underground Bus Stations

At underground bus stations the danger potential is slightly different compared to the underground car parks. Bus stations are characterized by a high number of people and are usually located at main traffic junctions, which offer transfer options to other buses or trains or are connected to shopping centers. In case of an evacuation, many people could be affected. There are currently several underground bus stations around the world, e. g. in Buenos Aires (Argentina), Stockholm (Sweden), Amsterdam (Netherlands), Berlin (Germany). Most likely the number of underground bus stations will continue to increase, as lack of space in major cities lead to new city concepts based on more underground infrastructure.

Because of environmental, political and social pressure, transport companies and operators are increasingly forced to convert to alternative drive systems, which also changes the spectrum of danger for bus stations. So far, there are no uniform concepts or recommendations for dealing with NEC in underground bus stations or the construction of underground bus stations.

Bus Depots

So far only a limited number of underground bus depots are in operation, for example in Fribourg (Switzerland), Stockholm (Sweden), Barcelona (Spain), El Monte (USA), Stavanger (Norway), Singapore (Singapore). Due to limited space in city centers and growing market share of NEC Buses the number of underground bus depots is expected to grow significantly.

Like underground garages, these are characterized by high fire loads and a high risk of fire spreading between the buses parked close to each other. Gas-powered buses further increase the risk of spreading due to excess gas blown off by the pressure relief valve. Without ignition, this can generate an explosive gas mixture in the environment. In connection with fire hazardous jet flame may occur. For operational reasons battery-powered buses need to be charged during parking in the depot. As described above the charging process embraces a significantly higher risk of fire. Evacuation is less critical, as in bus depots only trained staff is affected in case of fire. To detect fires immediately the detection systems must be well-matched to the hazards of the NEC.

Delivery zone

Delivery zones are increasingly being built underground for shopping centers. As many transport vehicles are powered with NEC, these enter the underground delivery zones and need to be considered in the safety regulations. Adequate detection systems and firefighting systems must be adapted to the new situation so that a safe evacuation of the usually small number of people is possible and damage to the infrastructure can be prevented. SUVEREN considers case studies in real delivery zones. The aim is to develop relevant recommendations for safety technology and to highlight the importance of the NEC with regard to the load of the transport vehicles.

Road Tunnel

Even if statistics show, that tunnels are the safest sections of the roads the situation in case of a fire may be very challenging, as survival conditions may deteriorate rapidly and tunnels provide limited options for escape. For this reason, the safety regulations for tunnels, although different in each country, are very high. New threats due to NEC are likely already covered by the design fires between 30 MW up to 100 MW considered for conventional fuels [RABT, 2016] and no additional restrictions for NEC vehicles are required. However, with increasing use of NEC, the transport of batteries and other hazardous materials will increase. Many regulations are set by the ADR (the European Agreement concerning the International Carriage of Dangerous Goods by Road), but some of these are contradictory and it must be examined to what extent NEC are dangerous for the users and the structure.

First, to investigate additional hazards from NEC following fire scenarios are defined independently from the above described case studies (underground facilities). Subsequent these reference scenarios are assigned to different cases (see Tables 1 to 3) concerning their potential risk and importance for each case. SUVEREN focusses on the risks of CNG (compressed natural gas, representing gaseous pressurized NEC), Li-Ion batteries (LIB) and composite materials (CM).

By defining these scenarios as a reference, it is possible to assign them to the above described underground facilities. Combining specific types and numbers of vehicles with the underground facilities allows to determine the main goals of investigations. Table 4 provides an overview about the summarized information and the five case studies that are investigated within SUREVEN.

Table 1. Compressed natural gas (CNG) Scenarios.

CNG 1:	A vehicle starts burning inside (i.e. due to technical failure in the motor department, cigarette, arson, etc.) → the fire spreads over the whole vehicle → a temperature triggered pressure relief device is activated → gas is released, and a jet fire occurs → surroundings are affected by the jet fire
CNG 2:	Due to a rupture or a pressure relief valve gas is released → a jet fire occurs after being ignited by an external ignition source → surroundings are affected by the jet fire
CNG 3:	A heat load from an external heat source (i.e. due to external vehicle fire) occurs → a temperature triggered pressure relief device is activated → gas is released, and a jet fire occurs → the burning behavior of the vehicle is influenced by the jet fire

Table 2. Li-Ion-Battery (LIB) Scenarios.

LIB 1:	A damage occurs inside the battery (thermal, mechanical or electrical failure) → the battery management system fails → a thermal runaway inside the battery arises → the vehicle starts burning
LIB 2:	A vehicle starts burning inside (i.e. technical failure in the motor department, cigarette, arson, etc.) → the thermal load induces a thermal runaway inside the battery → the vehicle starts burning
LIB 3	A heat load from an external heat source (i.e. external vehicle fire) occurs → parts of the vehicle ignite, and the vehicle starts burning → the heat load induces a thermal runaway inside the battery → the burning behavior of the vehicle is influenced by the battery fire

Table 3. Composite materials (CM) Scenarios.

CM 1:	A heat load from an external heat source (i.e. external vehicle fire) occurs → parts of the vehicle body start burning → the composite materials influence the heat release of the vehicle fire → surroundings are affected
CM 2:	A vehicle starts burning inside (i.e. technical failure in the motor department, cigarette, arson, etc.) → the composite materials influence the heat release of the vehicle fire → surroundings are affected

Table 4. Case studies and associated scenarios of SUVEREN.

Case Study (Underground Facility)	Main Goal of Investigations	Scenarios, No. of Vehicles	Measures to be evaluated
Car Park	• predict the fire spread between passenger cars • detect changes in smoke spread and releases of toxic gases • evaluate the influence of charging stations	6 passenger cars (2x3), 1 car starts burning: LIB 1, 3 CNG 1, 3 CM 1, 2	• examine the suitability of water mist systems • dedicated regulation for charging stations and/or charging areas in car parks • detection unintended gas releases in consequence of pressure relief failures • analyze early warning systems (eCall) for NEC-driven cars • examine additional smoke extraction systems
Bus Terminal	• predict the fire spread inside a bus • detect changes in smoke spread and additional gas releases	1 bus starts burning: LIB 1, 2 CNG 1, 2	• examine the suitability of water mist systems • adjustment of the PRD release direction • detect unintended gas releases in consequence of pressure relief failures • upgrading and adjustment of bus stations • examine additional smoke extraction systems
Bus Depots	• predict fire spread between several busses • evaluate the influence of charging stations • detect changes in smoke spread and additional gas releases	2 busses, 1 bus starts burning: LIB 1, 3 CNG 1, 3	• examine the suitability of water mist systems • regulation of charging stations and/or charging areas in bus depots • adjustment of the PRD release direction • upgrading and adjustment of parking spaces in bus depots • analyze early warning systems (eCall) for NEC-driven cars
Delivery Zone	• predict the fire spread inside a transporter/ truck • investigate the heat and smoke releases from transporters/ trucks	1 transporter/ truck starts burning: CNG 1 LIB 1, 2	• examine the suitability of water mist systems • consider the effect of fire detection devices inside battery packs • examine additional smoke extraction systems
Tunnel	• analytical estimation of HRR	1 truck loaded with batteries, 1 battery starts burning: LIB 1	• propose valve-controlled firefighting systems • consider the effect of fire detection devices inside battery packs

Figure 5.   Large-scale fire testing.

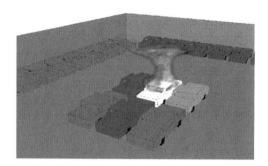

Figure 6.   Flame formation during the fire simulation of a car in an underground car park.

3.3   *Concept of large scale test programm*

To determine the properties of NEC in case of fire in connection to the relevant scenarios described before fire tests will be performed. As fire tests in general require severe effort and cause high costs a main goal of the project is to develop alternative methodologies. For this reason, an equivalent fire load with comparable properties shall be developed. The aim is to determine a cost-effective alternative to expensive real fire tests and to set a standard fire load for future tests.

At the end of the year 2018, fire tests will be carried out by the subcontractor IFAB in order to achieve the above-mentioned goals. Large Li-Ion batteries (30 kWh and 40 kWh) will be tested stand alone and in connection with active fire suppression systems (with and without water mist). Further tests with a gas tank set on fire will be executed to research the resulting jet flame and option for fire suppression. The fire tests are required to determine detailed parameters for CFD modelling and validation (see 3.4). In the respective test, a jet flame will be induced and sprayed with water mist to determine the interaction between the fine water droplets and the flame. The main goal is to repeat the results of the original fire tests with the equivalent fire load and to simulate the system with numerical calculations.

3.4   *Modelling of heat and species distribution and fire suppression*

A main task of SUVEREN is the quantification of the various risks and hazards from both new and conventional energy-driven vehicles and associated infrastructure (e.g. charging station) in underground facilities. As the targeted facilities are large and complex the analysis of the consequences of a NEC induced fire will be performed using computational fluid dynamics (CFD). CFD tools are widely used in both industry and research and proven to be capable of predicting the smoke and heat distribution in large buildings especially when fire testing is impossible or very cost intensive.

The individual assessment of every scenario (e.g. based on geometry, fire size and ventilation condition) enables a performance-based design approach. Further CFD modelling of SUVEREN will include the interaction of fire with active suppression and ventilation systems and consider toxic species other than standard fire products soot and carbon monoxide (CO). To improve the quality of the numerical calculations the models will be validated against the results of the fire tests.

4 OUTLOOK RESULTS

The research project SUVEREN (Duration 9/2017-8/2020) will deliver holistic safety concepts for underground, urban traffic areas, covering the challenges of both conventional and NEC fuels. This includes:

- Risk assessment methods including NEC
- Validated numerical models to identify hazards
- Recommendations for planning and equipment
- Development of appropriate mitigation technologies (active measures)

4.1 *Development of mitigation technologies*

In addition to determining the risk of NEC, one main objectives of SUVEREN is to develop and demonstrate the effectiveness of active measures to identify risks, increase the safety of users and reduce the damage. To achieve this, research is executed regarding the interaction between smoke gases and extinguishing agents as well as firefighting methods. For example, the effect of sprinklers or water mist extinguishing systems on battery fires is examined.

4.2 *Performance-based design*

For road tunnels the design and respective safety equipment (ventilation, fire protection) are defined by existing regulations (e. g. German RABT). The worst case scenario for these regulations are fires including heavy-goods-traffic (HGV) which are already dimensioned large enough to cover the impact of NEC in reasonable cases too. Consequently the use of NEC inside a road-tunnel equipped in accordance to current regulation will most likely not require additional safety equipment. Nevertheless, the rescue service must apply slightly different procedures and rescue strategies. Regarding road tunnels the situation differs considerably from underground car parks, where the fire risk increases due to NEC (e.g. charging stations for E-Mobility) and up to now prescriptive regulations apply, which may not cover the new requirements.

Car parks are equipped with different technical systems for guidance of the motorists, but also include firefighting equipment. Fire detection is mandatory in most countries; sprinkler systems in some countries. Other countries require sprinklers system only in the case that further constructions are was built above the car park. Hydrants and portable fire extinguishers are also mandatory in most countries. Car parks are typically divided into fire sections limiting the access of smoke & heat to larger areas using fire-rated barriers.

The findings of SUVEREN reveal that the materials used inside cars have changed and due to the increased use of combustible materials like plastics and composites, the fire load has changed for all current vehicles. Additionally, the fire load can vary a lot depending on car type, fuel (various NEC or conventional) and size.

To cover the wide scope of constraints a performance-based approach for dimensioning might be recommended for car parks following the success of introducing this method to assess tunnel projects. SUVEREN provides efficient tools and measure for a performance-based design approach: Risk assessment methods including NEC validated numerical models to identify hazards, appropriate and proofed mitigation technologies, and recommendations for planning and equipment.

4.3 *Guidelines*

Existing guidelines are based on conventional vehicles and do not cover changes due to increased use of NEC fuels or even changes in the fire loads of conventional vehicles due to new equipment and vehicle size or innovative materials like composites.

The results of SUVEREN are expected to be implemented in standards and regulations with reference to underground infrastructures. To offer concrete support to operators and designers of urban underground facilities a guideline with practical recommendations will be published and training programs will be offered.

5 CONCLUSIONS

The results available so far show that the increased use of NEC in underground urban infra-structures will not lead to a critical change of the safety level in general. Depending on the environment, the existing regulations and methodology cover the risks of NEC. Especially in case of road tunnels, regulations based on heavy goods vehicles (HGV) with conventional fuel and loads require little adjustments to the new situation. On the other hand, the situation in the other cases like underground car parks differs considerably, as the fire risk increases and prescriptive regulations do not cover the requirements. Here a performance-based approach for dimensioning will be investigated to embrace the large scope of constraints. SUVEREN provides efficient tools and measures to execute a performance-based approach: Risk assess-ment methods including NEC, validated numerical models to identify hazards, appropriate mitigation technologies, and recommendations for planning and equipment.

SUVEREN started in September 2017 and requires a duration of three years to perform the whole task. Large-scale test as an important base for the validation process will be executed end of 2018. Due to the early editorial deadline in August 2018 the given paper is limited to conceptual and preliminary results of the project. For recent details please consult: www.suve ren-nec.info

REFERENCES

Dutch Safety Board, 2012. Fire in a CNG Bus, Den Haag, Netherlands

Gehandler J. & Karlsson P. & Vylund L., 2017. Risks associated with alternative fuels in road tunnels and underground garages, SP – Technical Research Institute of Sweden, Report 2017:14, Sweden

RABT, 2016. Richtlinien für die Ausstattung und den Betrieb von Straßentunneln – RABT, Forschungs-gesellschaft für Straßen- und Verkehrswesen - Arbeitsgruppe Verkehrsführung und Verkehrssicherheit, Germany

Shell, 2015. Shell Flüssiggas-Studie, LPG als Energieträger und Kraftstoff Fakten, Trends und Perspektiven, Hamburg Germany

SUVEREN – Sicherheit in unterirdischen städtischen Verkehrsbereichen bei Einsatz neuer Energieträger, 2018, preliminary project reports, unpublished

Zhang Q. & Cunjin L. & Yuqing W., 2017. Analysis of Research and Development Trend of the Battery Technology in Electric Vehicle with the Perspective of Patent, Beijing China

Pictures credit by: 1: Yaorusheng – Shutterstock; 2: SUVEREN; 3: FOGTEC; IFAB; 4: Nerthuz - shutter stock.com; 5: IFAB; 6: FOGTEC

The project SUVEREN is funded by the German Federal Ministry of Education and Research under grant numbers 13N14391 to 13N14393. Responsible for the content of this publication are the authors.

Tunnels and Underground Cities: Engineering and Innovation meet Archaeology,
Architecture and Art, Volume 10: Strategic use of underground
space for resilient cities – Peila, Viggiani & Celestino (Eds)
© 2020 Taylor & Francis Group, London, ISBN 978-0-367-46878-1

Resilient city: The case of Bisagno diversion tunnel

G. Lunardi, G. Cassani, M. Gatti, B. Spigarelli & C.L. Zenti
Rocksoil S.p.A., Milan, Italy

ABSTRACT: In today's world of rapid environment changes, increasing urbanization and vulnerability, it is very crucial to embed the concept of resilience into the development planning of our cities in order to have a sustainable development. Planning a city without a robust resilience strategy to cope from disasters is equivalent to wasting resources and putting people, infrastructure, assets and economy at risk. Resilience is the ability of a system, community or society exposed to hazards to resist, absorb, accommodate, adapt to, transform and recover from the effects of a hazard in a timely and efficient manner. The Bisagno Valley has been historically affected by a number of flood events, which became particularly frequent in the last decade. This is mainly due to the urbanisation of the valley since the 1950s and the recent climate changes, with increased flash flood events. The paper describes the approach used on the Bisagno Diversion Tunnel Project Management, which involved stakeholder relationships and integration at different levels, institutional frameworks and partnerships amongst all urban stakeholders, particularly planner architects, engineers, disaster and risk reduction management specialists, private sector, and communities to address risk reduction and resilience in a holistic manner. Risk reduction and resilience building save lives, enhance social and economic development, and provide equitable, prosperous and sustainable urban development.

1 INTRODUCTION

The enormous potential for disaster resulting from natural hazards has been well known in Europe since the final decade of the 20th century. These issues were highlighted, particularly for urban areas, by the European Union conference which took place in October 1993 exploring the development of cohesive civil protection policies as part of the United Nations International Decade of Natural Disaster Reduction (Horlick-Jones et al., 1995). Horlick-Jones (1995) has stressed the need for better dialogue between researchers and practitioners of civil protection issues, a field that involves several different scientific, socio-economic, psychological, cultural and practical factors. Increasing population density and numbers of settlements in hazardous areas make disasters more frequent, severe and expensive (Petak, 1985; Drabek, 2004; Barredo, 2007; Castaldini & Ghinoi, 2009; Alberto et al., 2010), and as a result, the role of communication becomes more and more important (Horlick-Jones, 1995, Pearce, 2003; Arattano et al., 2010).

2 GENOA AREA

The city of Genoa, currently home to around 650 000 people, represents an Italian national case-study of the issue of geo-hydrological risk. In fact, during the last century the municipality of Genoa has been affected by recurring flood events and landslides that have caused heavy damage and casualties. In 2001, the geo-hydrological critical condition of the Bisagno catchment was defined by the Italian Civil Protection Agency as a "national emergency". The

Bisagno Stream flows through the most urbanized part of Genoa, with around 100 000 inhabitants as well as associated economic and industrial activity. Geo-hydrological risk mitigation in the Bisagno catchment area is therefore currently one of the most important civil protection objectives in Italy (Agenzia di Protezione Civile, 2001). The occurrence of very short hydrological runoff times makes accurate weather forecasts vital, with the time window during which potential intervention could take place in an emergency being very narrow.

The area of Genoa is characterized by a complex morphology determined by the Alpine–Apennine system which hosts relief extending from peaks between 1000 and 2000 m, rapidly descending towards the Ligurian Sea. The resulting hydrographic network consists of numerous steep and short watercourses that can attain a concentration time of less than an hour during floods (Figure 1).

The two most important catchments are the Polcevera stream which is the largest and the most populous basin (140 km^2), located west of the historic amphitheatre, and Bisagno stream (95 km^2) flowing immediately to the east. Important urban areas are also located in the plain coastal basins of Leiro stream at Voltri (27 km^2), of Varenna stream at Pegli (22 km^2) and Chiaravagna stream at Sestri Ponente (11 km^2). Following an increasingly widespread practice in Liguria, many of the Genoa river beds are culverted (covered), sometimes for long stretches, especially the reaches towards the mouth. In these new narrow spaces, roads, parking areas and, in some cases, even homes, have been built.

The municipality land, with its orogenic complexity, has peculiar features that make it unique geologically and hydrologically. The main characteristics are briefly listed in Figure 2. In the coastal plain the main streams feature Quaternary deposits that are now largely removed by anthropogenic actions (Comune di Genova, 1997; Giammarino et al., 2002). The mountain area at its back exhibits its major feature in the sector between Sestri and Voltaggio, a tectonic system that joins the westward Alpine units and eastward Apennine units (Corte- sogno & Haccard, 1984). It can be split into two units: one made up of ophiolites with corresponding meta- morphic sedimentary cover and one of calcareous dolomite. West of the Sestri-Voltaggio area, the units of the Voltri Group are found: a meta-ophiolite and metasedimentary complex representing the southernmost sector of the western Alps. From a lithological perspective, lime-shale, metabasite and ultramafic units are found here (Capponi et al., 1994). East of the Sestri-Voltaggio area, where the Polcevera Valley is located, shaly and shaly-limestone

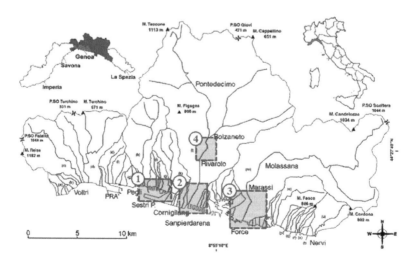

Figure 1. Main catchments and their streams, crossing Genoa in the stretch between the districts of Voltri (W) and Nervi (E): in the boxes (red dashed lines), from west to east: (1) Sestri Ponente and the mouth of Chiaravagna stream, (2) Cornigliano and the mouth of Polcevera stream, (3) Force and the mouth of Bisagno stream. In the centre of the figure, the red dashed box shows (4) Polcevera stream near Bolzaneto. (Faccini et al., 2015).

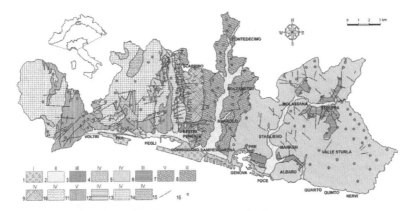

Figure 2. Geo-lithological sketch map with hydrogeological elements of Genoa Municipality. Legend: 1) embankment, dumps; 2) alluvial deposits; 3) stiff fissured clays; 4) ophiolitic conglomerates and breccias; 5) marly limestones with shale interlayers; 6) shales; 7) calcareous marls and silty shales; 8) shales with limestone interlayers; 9) pillows basalt and ophiolitic breccias, serpentinites; 10) dolomites, limestones, dolomitic limestones; 11) calcareous, micaceous and quartz-schists; 12) metabasalts and metagabbros; 13) serpentinites; 14) lherzolites; 15) fault, certain or covered; 16) main springs. Permeability classes: I) soils with variable permeability; II) permeable by porosity soil; III) waterproof formations; IV) permeable formations for cracks; V) medium-low permeability formation (Faccini et al., 2008).

flysch can be observed, whereas the shaly-marly fysch of Mt. Antola crops up on the left bank of Polcevera stream, reaching the eastern municipality boundary; locally it is broken up by the underlying shaly base complex or by the marly-shaly Pliocene lens above it, where the main tectonic directrixes are located (Limoncelli & Marini, 1969; Marini, 1981; Marini, 1998). The complex geological and tectonic configuration and the geographical position involve an extremely complex hydrologic setting that commonly features permeable rock masses next to less permeable masses.

2.1 The Bisagno Valley

The Bisagno stream rises near Scoffera Pass (675 m), The main stream has a total length of 25km, and its mouth is in Genoa city centre, east of the natural amphitheatre of the old town. The hydrographical network is characterised by particularly short streams within limited water-basins: the whole Bisagno catchment is < 100km², and maximum river length is < 20km. Therefore, concentration time during heavy rain-fall events is very short. The maximum elevation of the catchment is1034 m, and the average gradient is 31%; while 10% of the territory has a gradient of > 75% and only 5% has a gradient of < 10%. In the Bisagno Valley, 60% of the territory has a gradient between 35% and 75%, while almost 70% of the catchment is included between 0 and 500 m above sea level. Today the urban area covers 15% of the catchment, while 57% of the territory is wooded and this represents the main land use category for the higher part of the valley, which has almost entirely been abandoned since the end of the nineteenth century. The remaining part (28%) is composed by open area sand scattered cultivated fields on terraces.

The Bisagno valley is geologically characterised by marly limestone flysch of Mt. Antola (Upper Cretaceous) and related base complex of Montoggio shales. Ortovero Pliocenic clays feature only at the city centre. The main stream is incised to 11km from the mouth, while the flood plain is heavily urbanized. The maximum width of the floodplain is roughly 300 m before it reaches the main culvert, which encases the stream for the last 1.4 km as far as the stream mouth; while other minor culverts partially cover the stretch between the A12 motor way and Genoa football stadium. The whole basin is subject to intense erosion conditions affected by tectonic control. The reactivation of erosion processes is related to the last drop of the base level (Brancucci & Paliaga, 2005; Paliaga, 2015).

2.2 The largest flood

The city of Genoa (Liguria, Italy) and the Bisagno Valley are affected by frequent floods, often with loss of human lives. Historically characterized by high flood hazards, the Bisagno Valley was recently affected by a flood event on 9 October 2014, less than three years after the tragic flood event of 4 November 2011 when six people died, including two children. In the last 50 years, four destructive floods occurred in the Bisagno Valley, in addition to some other events that caused significant damage and economic losses.

The Bisagno Valley is characterized by climatic and landform features that have been making the flood events historically common in the area. However, recent climate change and land-use variations, including some major modifications of the catchment basin, have progressively determined a decrease of the concentration time and an increase of runoff, solid transport, and flood hazard. Consequently, in recent decades a growth in the number of flood events occurred, to the extent that the Bisagno today is a famous case study on an international scale. The three largest flood events in terms of intensity and ground effects which affected the Bisagno Valley in the last two centuries (Table 1): the flood of 25 October 1822, well documented by contemporary sources, the flood of 8 October 1970, undoubtedly the most tragic on record, and the very recent event of 9 October 2014.

In 1822, the Bisagno stream occupied the entire alluvial plain from the city walls to the hill of Albaro. Its course was braided with many small channels. No buildings were along the riverbed, apart from a leper hospital and the village of Borgo Pila. By1970 the situation was completely different because of the building of the railway (1868) and urban sprawl to the east. The Bisagno stream is embanked upstream of the station and completely covered downstream, parallel to the stream are two roads.

The plain is completely built up. The bridge of St. Agata is 70 m wide and only five arches remain of the original 28. Ponte Pila does not exist anymore, and the riverbed is only 50 m. The culvert of the Bisagno stream was decided in the twentieth century: it was calculated by a group of three engineers on commission by the Genoa municipality in 1908 (Inglese et al., 1909):using definitely modest hourly values of precipitation, they calculated that the maximum flow rate of Bisagno could not exceed 500 m^3/s. Based on the underestimated data of 1908, a large project of urban development was carried out during the fascist period, and a large square was built that could celebrate the regime. In 2014 urbanization is complete; the embankments are characterized by roads, warehouses, and factories; while the alluvial plain and the sides of the valley are heavily built up with residential buildings. New culverts were built downstream of the cemetery of Staglieno at Genova Est motorway exit and near the football stadium. St. Agata bridge was repeatedly damaged and almost destroyed during the floods of 1970 and 1992. Of the ancient bridge, only two arches remain, located upstream of the final culvert, which is 50 m wide.

Table 1. Largest flood events of the Bisagno stream.

Storm event day	Rainfall event	Discharge	Flood event	Damage losses and other damage	Storm-related fatalities
1822-10-25	812 mm/24 h in the lower Bisagno catchment (Marassi)	1200 m^3/s	Regular flood. 15 h of violent rainfall a 3 h peak between 10 a.m. and 1 p.m	Many streams flooded: two bridges on the Bisagno collapsed. Mud and water reached the second floor of the houses at Foce district. Serious damage to shops, farms, factories and the public aqueduct. Estimated damage around half a million of Savoy liras. Not quantifiable by the historical sources	Un-know
1970-10-08	453 mm/24 h (PCAR) 394 mm/24 h (RGHI)	950 m^3/s end culvert underpressure	Regular flood. Flood of the Bisagno, Torbido, Geirato, Veilino, Fereggiano and Mermi streams	55 million Euros equivalent 1000 people left homeles 50,000 people without jobs 75 million Euros equivalent 250 people left homeless	10 fatalities only in the Bisagno valley, 44 in the Genoa metropolitan area No reports of fatalities in the Bisagno basin. 2 fatalities in the close Sturla basin 6 fatalities
2014-10-09	141 mm/1 h (GEGR). 401/24 h (GEGR)	1000 m^3/s final culvert underpressure	Flash flood. Flood of the Bisagno e Fereggiano streams	300 million Euros. 250 people left homeless	1 fatality

(a)   (b)

Figure 3.   Inundations in Genoa October 1970 Bisagno overflowing: a) in the zone downstream the covering near the Genoa Brignole railway station; b) via Canevari.

3   RESILIENT CITY

Resilience is "the ability of a system, community or society exposed to hazards to resist, absorb, accommodate, adapt to, transform and recover from the effects of a hazard in a timely and efficient manner." There is no one-size-fits-all solution to achieve resilience. Local government actors will determine how these actions apply to their own contexts and capacities. In the urban setting, risk management is an essential part of building resilience.

3.1   *Priorities for action*

The substantial reduction of disaster risk and losses in lives, livelihoods and health and in the economic, physical, social, cultural and environmental assets of persons, businesses, communities and countries could be achieved following priorities for action:

- *Understanding disaster risk*:
  Disaster risk management should be based on an understanding of disaster risk in all its dimensions of vulnerability, capacity, exposure of persons and assets, hazard characteristics and the environment. Such knowledge can be used for risk assessment, prevention, mitigation, preparedness and response.
- *Strengthening disaster risk governance to manage disaster risk:*
  Disaster risk governance at the national, regional and global levels is very important for prevention, mitigation, preparedness, response, recovery, and rehabilitation. It fosters collaboration and partnership.
- *Enhancing disaster preparedness for effective response and to "Build Back Better" in recovery, rehabilitation and reconstruction:*
  The growth of disaster risk means that there is a need to strengthen disaster preparedness for response, take action in the anticipation of events, and ensure capacities are in place for effective response and recovery at all levels. The recovery, rehabilitation and reconstruction phase is a critical opportunity to build back better, including through the integrating of disaster risk reduction into development measures.

3.2   *The Ten Essentials for Making Cities Disaster Resilient*

An effective risk reduction could be achieved by applying the Ten essential (Figure 4) defined by United Nation through the Sendai Framework for Disaster Risk Reduction

- *Organize for disaster resilience*. Put in place an organizational structure with strong leadership and clarity of coordination and responsibilities. Establish Disaster Risk Reduction as a key consideration throughout the City Vision or Strategic Plan.

Table 2.   City-Action Plan.

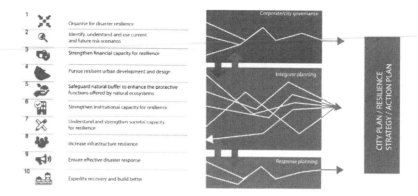

- *Identify, understand, and use current and future risk scenarios.* Maintain up-to-date data on hazards and vulnerabilities. Prepare risk assessments based on participatory processes and use these as the basis for urban development of the city and its long-term planning goals.
- *Strengthen financial capacity for resilience.* Prepare a financial plan by understanding and assessing the significant economic impacts of disasters. Identify and develop financial mechanisms to support resilience activities.
- *Pursue resilient urban development and design.* Carry out risk-informed urban planning and development based on up-to-date risk assessments with particular focus on vulnerable populations. Apply and enforce realistic, risk compliant building regulations.
- *Safeguard natural buffers to enhance the protective functions offered by natural ecosystems.* Identify, protect and monitor natural ecosystems within and outside the city geography and enhance their use for risk reduction.
- *Strengthen institutional capacity for resilience.* Understand institutional capacity for risk reduction including those of governmental organizations; private sector; academia, professional and civil society organizations, to help detect and strengthen gaps in resilience capacity.
- *Understand and strengthen societal capacity for resilience.* Identify and strengthen social connectedness and culture of mutual help through community and government initiatives and multimedia channels of communication.
- *Increase infrastructure resilience.* Develop a strategy for the protection, update and maintenance of critical infrastructure. Develop risk mitigating infrastructure where needed.
- *Ensure effective preparedness and disaster response.* Create and regularly update preparedness plans, connect with early warning systems and increase emergency and management capacities. 10. After any disaster, ensure that the needs of the affected population are placed at the centre of reconstruction, with support for them and their community organisations to design and help implement responses, including rebuilding homes and livelihoods.
- *Expedite recovery and build back better.* Establish post-disaster recovery, rehabilitation, and reconstruction strategies that are aligned with long-term planning and providing an improved city environment.

4   THE FLOOD PROTECTION WORKS OF GENOA: THE DIVERSION OF THE BISAGNO RIVER

4.1   *Introduction*

The diversion tunnel on the Bisagno River has been planned for flood protection in the city of Genoa. The first project was presented and approved in 2007, but despite the technical approval of the Superior Council of the Public Works, Ministry of the Public Works and Infrastructures, the works didn't start pending appropriate budget.

In November 2011 the city of Genoa was hit by a devastating flooding event, causing several deaths and strong damages to infrastructures, buildings, private and public goods. The most disastrous effects were due to the flooding of Bisagno River and Fereggiano creek. The Fereggiano estimated peak flow was in the range of 140 m³/s, corresponding to a discharge flow with an associated return period T larger than 200 years.

After this event, in 2013 the City of Genoa granted the first allotment contract (I Lot – 50 million Euro allocated to solve the major hydraulic problems), that includes the safety measures for the Fereggiano creek and for the Rovare and Noce streams: among those, the Fereggiano tunnel to the sea outfall (diameter: 5.2 m; length: about 3700 m) and the catchment works by the Rovare and Noce streams (Ferrari et al. 2014). In 2017, according to the Government's Found for "Safety Italy", the detailed design for the Bisagno Diversion Tunnel was assigned, by the Special Commissioner, to a group of Designers, including three company: Rocksoil, as chief company, Hydrodata and Art, plus some specialists: Cangiano, De Sanctis, Gallo and Giomi.

4.2 *Project overview and construction systems*

The detailed design of II lot, strictly related to the Bisagno diversion tunnel and related civil works, started last August and now the project, after the exam of the Supreme Council of the Public Works, has been subjected to the study of environmental impact (VIA procedure). As for Lot I a revision of the final design, approved in 2007, has been necessary. Figure 4 shows the general overview of the project: the aim of the project is to intercept the flood flows of the Bisagno river, at the Sciorba sports complex, and discharge them into the sea through a diversion tunnel with a total length of about 6,500 m.

The project is divided into 3 main parts: upstream section, central section and downstream section. The main element of the *upstream section* (Figure 5) is the intake system, composed by the Bisagno river barrage and the lateral spillway; at this point the Bisagno's flow is splitted between the original culvert (renwed in the underground strech from Brignole Station to the sea outfall) and the diversion tunnel. The upstream side of the diversion tunnel will be bored by conventional excavation method (645 m) starting from the area of the former kennel of Genoa, located near the Sciorba stream. In this area will be located the working site and an adit tunnel will start to reach the diversion alignment by means a conventional tunnel, 398 m long. The adit will intersect the diversion tunnel into a big cavern, which will be used to assemble the TBM and to launch the mechanised tunnel to south towords the sea outfall; the mechanised bored section of the diversion tunnel is 5785 m long with a diameter of 10.7 m. Starting from

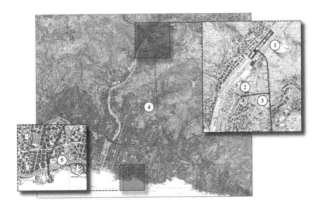

Figure 4.   II lot project overview: (1) Intake system with Bisagno River Barrage and lateral Spillway; (2) Sciorba Intake; (3) Service tunnel and cavern; (4) Diversion Tunnel; (5) Link to the sea outfall.

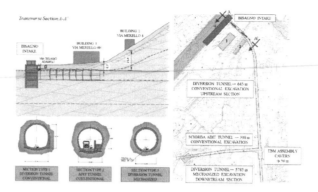

Figure 5.   Bisagno diversion tunnel: upstream section.

the cavern the north section, up to the Bisagno river, will be excavated too, by conventional method. The last part of this upstream section will underpass Merello and Adamoli Streets, where some residential buildings are located, together with the football and swimming sports center. The adit will be used to spillway the Sciora strem too. The final stretch of the Sciorba stream flows in an urban area, at the right of the Bisagno River and it is located about 650 m downstream of the Bisagno intake. The Sciorba stream flows underground in correspondence with the planned construction site area, because in the past the Sciorba stream has been channeled and trimmed to allow the construction of a series of service buildings in that area.

The construction of the Bisagno diversion tunnel offered the opportunity to realize the spillway of the Sciorba in order to reduce the flow rates in transit along the underground section, which in the past caused flooding due to hydraulic inadequacy of its culvert. The figure 5 shows the different section type designed for the described tunnel stretches, considering the specific tunnel function and the applied excavation method. The construction system for these strectches provides the use of the "drill&blast" method, placing a prelining after excavation, composed by steel bolts or ribs and shotcrete layers and casting in situ the final lining. For the last stretch, near the Bisagno river, where low overburden is present, grouting treatments will be executed before excavation and low vibration system will be used to reduce noise and interference with the buildings ("super-wedge" and "smooth blasting" systems).

The diversion tunnel located downstream to the cavern will be excavated by TBM up to the sea outfall and it is identified as the *central section* of the project. The excavation will be executed by an open "hard-rock" TBM, equipped by drilling systems, able to execute probe drilling, forepoling and grouting activities in advance of the core-face and radial bolting for the excavated cavity. The TBM will advance by grippers and the prelining will be mainly composed by radial bolts and steel wire-mesh; the shotcrete layer will be placed by a robot system about 40–50 m far from the face, in fractured rock-mass the shotcrete could be placed immediately back to the face by a manual system. The back-up of the TBM will place the invert segment, equipped with tracks to manage the following formworks to cast in situ the final lining. Waterproofing will be placed, together with drainages, located radially the final lining, to dewater the groundwater level up to a water-pressure to be supported by the final lining.

The *downstream section* defines the sea outfall system. The final section of the spillway (50 m) will underpass the promenade of Corso Italia and will be bored by conventional excavation method from the arched structure that supports the promenade.

The excavation will start from the Corso Italia beach and will be executed by grouting treatments to create the "arch effect" around the excavation profile and waterproof the fractured rock-mass: the sea-level intersects the tunnel section. In Figure 6 (left) a longitudinal profile of the tunnel is shown, with the grouting activities in green, and (right) the section of the diversion tunnel under the arches of Corso Italia. Note on the left the section of the Fereggiano

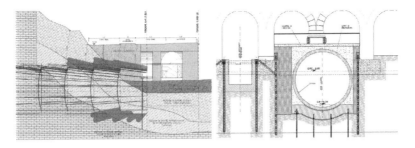

Figure 6. Bisagno diversion tunnel: downstream section. Longitudinal and transverse sections.

channel, which reaches the sea in the same point of the Bisagno diversion by means of a common portal.

The TBM coming from upstream will be stopped at the arrival shaft, located in the beach, passing throught the conventional downstream section. The barrier delimiting the construction will be conver by information panels which will update the public about the state of work (Figure 7).

4.3 *Hydraulic dataset*

The Bisagno diversion has been designed in order to take a flow of 450 m³/sec, so that the flow in the final section, near to the mouth of the Bisagno, does not exceed 850 m³/sec (considering the whole flow, with an associated return period T of 200 years, is 1300 m³/sec taking into account the rainfall statistics).

The tunnel capacity has been checked with reference to a flow of 460 m³/sec, considering the flow derived from the Sciorba stream, equal to 8 m³/sec. The capacity has been verified by both numerical analyses and physical modelling in real tests. The degree of filling, considering a Strickler coefficient equal to 65 m$^{1/3}$/s, is always less that 75%; it is not foreseen a "in pressure" behavior of the tunnel. The inset provides a slide with a Creager profile and a dissipation tank to introduce the water into the tunnel diversion. The flow diverted into the tunnel is regulated by a barrier with adjustable gates; in this section the pavement of the river is coated with stones and new banks will be made; upstream a selected bride is located, to stop any solid transport. Near the intake, a building will be constructed with the operational equipment and with a control room.

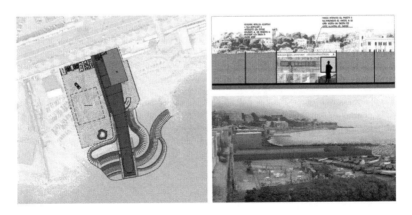

Figure 7. Bisagno diversion tunnel: downstream section. Plan and area overview.

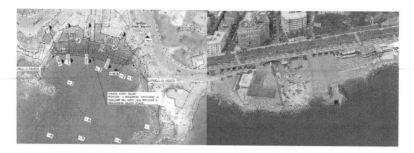

Figure 8. a) Nourishment of Quarto Bolivar beach. b) Outfall area of the Bisagno diversion.

4.4 Geological and Environmental aspects

The alignment of the Bisagno diversion is mainly affected by limestones and marls of the Formation of the "Flysch del Monte Antola", with overburdens ranging from 10–20 m up to 280 m. Just a strict section, about 200 m long, is interested by shale and claystones of the Formation of the "Argilliti di Montoggio" with overburden of 40–45 m.

The upstream section, where the diversion tunnel is connected with the Bisagno river, passes throught alluvional deposits too, composed by gravels and sands, as well as the downstream section which is interested by the fractured limestones rock-mass and by the sandy beaches of Corso Italia. The characteristics of the limestones are suitable to use the materials resulting from the excavation to produce shotecrete and concrete for the tunnel constructions. Furthermore the excavated material, appropriately shattered and washed, will be used for the nourishment of the beaches along the coasts of Genoa. Properly projects have been prepared to manage the activities for the beaches, transforming the usual problems of the relocations of the excavated materials in a social opportunity for the community of Genoa.

ACKNOWLEDGEMENTS

The authors wish to thank the colleagues of the Design Group and the officials of the Municipality of Genoa, Stefano Pinasco, and of the Liguria region, Luca Berruti, for the support during the preparation of the design stage and of the approval process.

REFERENCE

Alberto, W., Giardino, M., Martinotti, G., & Tiranti, D., 2008. Geomorphological hazards related to deep dissolution phenomena in the Western Italian Alps: Distribution, assessment and interaction with human activities, *Eng. Geol.*, 99, 147–159.
Arattano, M., Conte, R., Franzi, L., Giordan, D., Lazzari, A., & Luino, F., 2010. Risk management on an alluvial fan: a case study of the 2008 debris-flow event at Villar Pellice (Piedmont, N–W Italy), *Nat. Hazards Earth Syst. Sci.*, 10, 999–1008.
Barredo, J. I. 2007. Major flood disasters in Europe: 1950–2005, *Nat. Hazards*, 42, 125–148.
Brandolini, P., Cevasco, A., Firpo, M., Robbiano, A., & Sacchini, A. 2012. Geo-hydrological risk management for civil protection purposes in the urban area of Genoa (Liguria, NW Italy), *Nat. Hazards Earth Syst. Sci.*, 12: 943–959.
Castaldini, D. & Ghinoi, A. 2009. Studio della pericolosità geomorfologica in aree produttive del bacino montano del fiume Panaro (Appennino Settentrionale), *Ambiente geomorfologico e attività dell'uomo, risorse, rischi, impatti,* edited by: Agnesi, V., Memorie della Società Geografica Italiana, 87, Roma, 87–98.
Comune di Genova, 1997. Atlante Cartografico Geologico del territorio del Comune di Genova, *Cartografia geologica, geomorfologica ed idrogeologica in scala 1:10.000,* System Cart S.r.l.
Faccini, F & Vassalli, G. 2008. Geological hazard and abandoned waste management: The case of the Genoan Municipality (Liguria, Italy), *Geoingegneria Ambientale e Mineraria,* 124: 55–61.

Faccini, F., Luino, F., Sacchini, A., Turconi, L., and De Graff, J. V. 2015. Geohydrological hazards and urban development in the Mediterranean area: an example from Genoa (Liguria, Italy), *Nat. Hazards Earth Syst. Sci.*, 15: 2631–2652.

Ferrari, M., Belicchi, M., Cerlini, D., Majone, U., Venturini, S., Marchi, A., Galli, A., Galli, U. & Pinasco, S. 2014. The Bisagno River Diversion. *In: River* Flow *2014*, 1757–1765. London: CRC Press – Schleiss A.J., De Cesare G., Franca M.J., Pfister M. (eds).

Horlick-Jones, T., Amendola, A., & Casale, R. (Eds.): *Natural risk and Civil Protection*, E &FN Spon, London, 1995.

Pearce, L., 2003. Disaster Management and Community Planning, and Public Participation: How to Achieve Sustainable Hazard Mitigation, *Nat. Hazards*, 28, 211–228.

Petak, W. J. 1985. Emergency Management: A Challenge for Public Administration, *Public Administration Review*, 45, Special Issue: Emergency Management: A Challenge for Public Administration, 3–7.

Rossi, L.: *Innovation in impacts forecasting and multi-hazard risk assessment*, Cima Research Foundation partner of Disaster Risk Management Knowledge Centre belonging to the European Commission https://drmkc.jrc.ec.europa.eu

Tunnels and Underground Cities: Engineering and Innovation meet Archaeology, Architecture and Art, Volume 10: Strategic use of underground space for resilient cities – Peila, Viggiani & Celestino (Eds)
© 2020 Taylor & Francis Group, London, ISBN 978-0-367-46878-1

Genoa underground: Solidity, usefulness and beauty of the Grazie Bassa Tunnel during 100 year service life

C. Panariello, H. Mohamed Dahir & V. Molinari
Italferr S.p.a, Rome, Italy

ABSTRACT: The Grazie Bassa tunnel is one of the first railway infrastructural works (1920) carried out underground in correspondence of the oldest part of the territory of the city of Genoa, a densely urbanized area with a predominantly vertical development going up from the ancient port to the residential districts of Carignano and Castelletto.

The paper aims to summarize the transformations, both in constructive terms ("firmitas"-solidity) and in function ("utilitas"- usefulness), that characterize the history of the tunnel from its construction to today, focusing on what has happened in the last ten years in which long stretches of the old tunnel have been deeply transformed to become integrated parts of the new metro and railway tunnels. The rehabilitation works have been also a great opportunity to analyze old construction methods and to measure and analyze stress and resistance of the old masonry lining. An idea for a reuse of part of the tunnel in decorative way ("venustas" – beauty) is also mentioned.

1 INTRODUCTION

The Grazie Bassa Tunnel was one of the earliest pieces of underground railway infrastructure to be excavated below the oldest part of the city of Genoa, a densely urbanised territory that extends upwards from the historic port (the Porto Antico), following the slopes of the hills until it reaches the residential areas of Carignano and Castelletto.

The urban character of the most historic part of Genoa comes from the steep alleyways, known in the local dialect as the "crêuze" that form a complex system of pedestrian routes, mainly above the ground, and from the network of the most important railway lines that were excavated below the ground. The crêuze and the railways are both fully integrated by lifts and funicular railways.

Based on the Vitruvian concepts of firmitas (solidity) and utilitas (functional usefulness), this paper summarises the transformations that characterise the history of the Grazie Bassa Tunnel in terms of its construction, from when it was first built up until today, with a particular focus on developments over the past ten years, when modifications were carried out that began from the tunnel portal at the Brignole railway station and ended at the point where the tunnel passes underneath the older Traversata Vecchia and Traversata Nuova Tunnels.

We also offer possible other elements which, because of their pure venustas (aesthetic delight), could be a stimulus for further transformations.

2 THE ORIGINAL CONSTRUCTION

The construction of the twin-bore Grazie Bassa Tunnel began in 1910 as the completion of the complex Genoa railway system, adding an underground connection for transporting freight

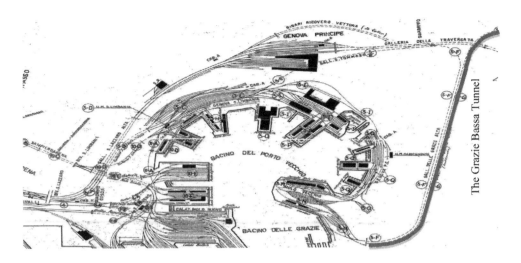

Figure 1.   City of Genoa early 1900s – Diagram of railways.

between the Porto Antico and the stations at Genova Piazza Principe in one direction and Genova Brignole in the other.

The works consisted of excavating a pair of tunnels, denominated "delle Grazie", beginning east of the Molo Vecchio (the historic pier) at the harbour and running side by side for an initial stretch of approximately 800m before separating at a point corresponding to Piazza De Ferrari above.

One of the tunnels then deviated eastwards, terminating at Genova Brignole railway Station and enabling freight trains to run and from eastern Liguria.

The excavation passed through a marly-calcareous formation known as the" Monte Antola" flysch. Due to the characteristics of this rock mass, the excavation was carried out manually with the help of explosives.

The tunnel was lined with solid clay bricks for the flanks and the crown, with blocks of limestone for the abutments. The voids left by the irregularities of the excavation in relation to the lining were partly filled using the excavated material.

The cross-section of the tunnel had a maximum width at the abutments of approximately 4.80 m, a radius of 2.10 m at the crown, and a height from the finished floor of approximately 5.50 m.

The tunnel entered service in 1922 but was taken out of use in 1939 when rail freight transport was relocated from the Porto Antico to the marshalling yards at the west of the port.

3   THE WARTIME PERIOD

During the Second World War the Grazie Bassa Tunnel was in public use as an underground refuge from allied bombing raids.

On 23 October 1942, yet another of the many alarms sounded (this one later turned out to have been unnecessary) and the local residents hurried to the stairs leading down from the access point near Porta Soprana, where there were approximately 150 steps.

There was a panic and some people lost their balance, dragging others along with them, with devastating effects. The rescuers counted more than 350 victims.

Figure 2. Grazie Tunnel – Metro reused.

4 THE FIRST TRANSFORMATION: THE METRO SYSTEM

The Grazie Bassa Tunnel was "rediscovered" in the 1980s when a study of mobility and traffic in the city led to a decision by the municipality to transfer passenger movements from its surface tramway system, which by then had become inadequate, to a new underground Metro system.

The Genoa Metro between the stations at De Ferrari and Brignole reuses approximately 200 m of the existing Grazie Bassa Tunnel, between the above-ground locations of Via della Marina and Via Roma.

That decision was taken partly for economic reasons and partly because in that way, the Genoa Metro would have a lower impact on important existing buildings such as the Teatro Carlo Felice and the Galleria Mazzini.

But it meant that a lower profile had to be adopted for this section, which could only to take Metro carriages of lower capacity (width 2.20 m). The only construction works that were necessary, consisted mainly of consolidating the existing tunnel lining.

5 THE MODIFICATIONS BETWEEN 2008 AND 2018 – THE GENOA RAILWAY INTERCHANGE SYSTEM

5.1 *2008: Design and tender bid contest for the works*

As part of a strategic infrastructure development programme approved by the Interministerial Committee for Economic Planning, in 2003 the "Ferrovie dello Stato" Group (the Italian State Railways) began work on the Detailed Design to upgrade Genoa's railway interchange system.

Two new tracks had to be built in tunnel between Brignole and Principe stations, to give better separation between national/regional traffic and the urban services by increasing the number of tracks from 4 to 6. Therefore a Detailed Design was developed in 2008 by Italferr (Pigorini et al. 2010).

The two new single-track tunnels for the new lines were denominated the New San Tomaso Tunnel and the New Colombo Tunnel.

In 2008 a tender bid contest was held to construct these tunnels, each of which had an overall length of approximately 1500 m.

Both tunnels begin from the tunnel entrance at Brignole Station and connect to the existing San Tomaso and Colombo tunnels respectively; these are short connecting tunnels that link the below-ground station at Principe Sotterranea with the tunnels denominated Traversata Vecchia and Traversata Nuova.

Beginning from the existing tunnel entrance at Brignole, the first 850 metres of the New San Tomaso tunnel are superimposed on the alignment of the earlier Grazie Bassa Tunnel.

Unlike the work carried out for the section of this existing tunnel that was taken over by the Metro, in this case the existing inner cross-section was not large enough to permit the transit of trains. The existing tunnel lining was therefore improved with consolidation works and after enlarged by excavation. At the most critical points (where there was interference with buildings above and/or with existing railway tunnel in operation), the existing tunnel was completely filled with lightweight concrete.

The Traversata Vecchia and Traversata Nuova tunnels, which were in service, connected the above and below-ground railway stations at Principe Station with the railway station at Brignole.

Construction of the parts of the new San Tomaso tunnel that had to cross underneath the Traversata Vecchia and Traversata Nuova was particularly critical, considering in particular the reduced overburden, which at some points is less than 1 meter, between the top of the existing tunnel (as widened) and the floor that was to take the track bed. So for that specific situation a dedicated tunnel excavation sections were designed consisting of installing a stiff pre-supports (large diameter forepoling steel pipes) to the crown and a stiffer pre-lining than what is used in more average conditions, consisting of "strong" steel ribs set at closer centres; this was all associated with a detailed monitoring system to check the excavation advancement (convergences, tunnel face outcrop, stress measurements of prelining) and the stress/strain condition of the final linings (strain and vibration monitoring of the tunnel abutments).

5.2 2010 – 2011 Modification of the Brignole tunnels portal

These upgrading works to the Genoa railway system began in August 2010 with the very complex demolition of an existing building that stood very close to the portal of the Grazie Bassa tunnel.

This building consisted of two blocks each 27m high with a total footprint area of 1400 m², plus an interconnecting block, giving a total volume of 38,000 m³.

Figure 3.   The former ARPAL building to be demolished.

Figure 4.   The former ARPAL during the demolition.

The different stages of the demolition process were conditioned by interferences with the operating railway and the surrounding city streets.

Taking into account the significant daily traffic flows in the streets around the building (in Via Montesano and Via Gropallo) the Genoa Municipality issued a concession permitting those streets to be temporarily occupied for seven consecutive days during a period when there was less traffic (the end of the school year) in order for the demolition site to be made ready.

In view of the limited time available, the Director of Works proposed that the contractor should modify the Detailed Design (demolition floor by floor using a fixed crane) by using demolition equipment that was to be installed at the beginning of the process, both at high level using the closed streets, and at low level using the site access lane, whilst also taking steps to avoid any restrictions to the operation of railway.

The following equipment was used for the demolitions:

For the demolition itself: 3 no. excavators with demolition grabs on telescopic arms;

For dust abatement: 1 no. ready-mix concrete vehicle equipped with a pump on a telescopic arm for localised spraying, plus 2 no. fog cannons;

As protection from falling rubble during demolition: 2 no. reinforced PVC sheets supported by mobile cranes.

After a delicate stage during which asbestos-containing materials were removed, demolition began in September and was completed in approximately one month without any impact on the operation of the adjacent railway.

With the demolition completed but before construction could begin, an interference at the portal of the new San Tomaso and Colombo tunnels had to be resolved between the micro-piles for the retaining wall and an underground 132 kV high-voltage electrical cable. This proved particularly complex due to the existence of a great many other below-ground services, the limited space available, and the need to carry out investigations to locate any unexploded bombs that had been there since the Second World War.

Construction of the tunnel portal consisted of the following works:

*   Filling 80 m of the Grazie Bassa Tunnel with lightweight concrete;
*   Constructing a micropiles retaining wall;
*   Demolition/excavation works at the existing masonry tunnel portal
*   Constructing a reinforced concrete retaining wall including the entrances of the new tunnels

The micropiles for the "Berlin type" retaining wall, were 19.50m long with a diameter of 150mm.

The cross-section of the foundation of the reinforced concrete wall is 3.80 * 1.40m and the wall itself is 15 m high with a cross-section varying from 1.95m to 0.45m.

The characteristics of the most important structural materials used were:

*   S275/S355 steel for pipes and profiles;
*   B450C steel for reinforcement;
*   Rck > 30 MPa concrete.

Figure 5. Brignole tunnel portal: searching for archaeological remains.

The Detailed Design provided the demolition of a stretch of existing masonry retaining wall parallel to the railway by the extension of the provisional retaining wall. This required occupying a siding for approximately three months on which to construct a gravity gabion wall needed for implementing all the construction stages.

To eliminate any interference with railway operation during these works, a modification to the provisional and definitive retaining works was identified without the demolition existing wall parallel to the railway. So, it was constructed a reinforced concrete wall, which were adequately anchored and tied back to the pre-existing masonry supporting works.

The demolitions and excavations were carried out under the supervision of the Archaeological Superintendence because it was believed that remains would be found of the Convent of Nostra Signora del Rifugio (Our Lady of Refuge – Brignoline nuns).

In fact the excavations brought to light masonry that had been built before the convent; analysis of the mortars dated it to the mediaeval period and attributed it to palaces that had once belonged to the Grimaldi-Bacelli and Squarciafico families.

Figure 6. Tunnel portal Brignole side – demolition of the existing portal.

Figure 7.   The start of the excavations to widen the Grazie Tunnel.

The construction work at the tunnel portal, and the archaeological investigations being carried out at the same time, were appropriately coordinated and not only did not create any difficulties but enriched the works by adding greater scientific significance to them.

The stability of the temporary support works was put to the test, with positive outcomes, by a flood that hit Genoa in November 2011.

On that occasion the monitoring system did not register any significant movements and the watertightness of the shotcrete lining was confirmed visually.

When the connections between the retaining wall and the tunnel portal were completed in March 2012, excavation began to widen the Grazie Bassa Tunnel.

5.3   2012 – 2016: the excavation works to widen the Grazie Bassa Tunnel

During the advancement of the excavation to widen the Grazie Bassa Tunnel, using a roadheader, no particular critical points were encountered either in relation to the behaviour of the rock mass and the pre-lining works, nor any induced deformations in the buildings above.

This fully confirmed the original design hypotheses, which had predicted that there would be no damage whatever to any existing buildings, including buildings of architectural importance for which there was only thin overburden, such as those directly above the tunnel entrance.

Specifically, in fact, the monitoring results detected only millimetric values of convergence and settlement at the surface, and these were well below the warning thresholds (Pedone et al. 2014).

Similarly, the tensile state values of the buildings that were specifically monitored were also very far away from the warning thresholds the designers had indicated.

The only critical factor that had a conditioning affect on the regular progress of the works was the constant noise generated by the excavation work, particularly during excavation of the inverted arch, which was carried out using an excavator-mounted hydraulic hammer.

Although the noise emission limits within the dwellings above were not significantly exceeded, the municipal administration imposed a limit on night-time working, which meant that a continuous 24-hour work cycle was not possible.

The excavations to widen and lower the floor levels of the Traversata Vecchia and Traversata Nuova Tunnels were carried out during the summer months of 2015 and 2016.

In fact, differently from the provisions of the Detailed Design, the railway operator (RFI), preferred for these delicate implementation stages to be carried out when no trains were passing. For that reason, specific measures were adopted to modify train operation, including a suspension of railway traffic in the tunnel for approximately two months during summer holiday (the least busy time of year).

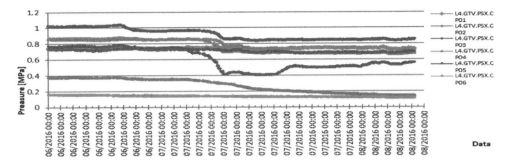

Figure 8. Extract from the monitoring report (flat jacks pressure recorded in GTV).

In fact although it necessarily introduced a stop to the excavation work because the advancement of the excavation face was out of phase, after the lowering of the second tunnel floor level was completed in relation to the restart of railway operation, this solution made it possible to avoid installing the temporary track support works that had been provided for in the Detailed Design, and also made monitoring more convenient of the other tunnels with which there were interferences, eliminating the need to continuously verify the track.

Again, in this case, the recorded levels remained well below the warning thresholds set out in the Detailed Design (Pigorini et al. 2016).

5.4 2016 – 2018: the intermediate stage; no particular transformations carried out

After completion of the tunnel final lining under the Traversata Vecchia Tunnel in August 2016, the works to the Grazie Bassa Tunnel were suspended as a result of early termination of the contractual relationship between the Client and the Contractor.

Preparation of the documentation required for the new tender bidding process was performed in 2017.

The build contract to complete the new San Tomaso and Colombo Tunnels, compliant with the currently applicable code (Legislative Decree 50/2016) was signed in March 2018.

The reinforced concrete final lining is currently being applied to the parts of the tunnel already excavated.

It is expected that construction of the escape tunnel in Piazza Corvetto will begin in the first months of 2019.

The contractual works programme provides for the transformation of the Grazie Bassa Tunnel to be completed within 2020.

Figure 9. Excavation to widen the Grazie Bassa Tunnel – cross-sections strengthened by filling with lightweight concrete.

6   CONCLUSIONS

As a summary description of the transformations of the Grazie Bassa Tunnel, seen from the particular point of view of the Direction of Works, this article has been an opportunity to appreciate the construction techniques that were used in the early 1900s and how they have evolved since then.

Stimulated by the reports that were presented at a recent national tunnelling conference, we have attempted to assess those transformations by making reference to a simplified version of "De Architettura", a treatise written by Vitruvius sometime around 15 BC, as summarised later by the French architect Claude Perrault in the 17th century, according to whom any work of architecture must satisfy the *triade vitruviana*, i.e. must meet the three criteria of:

- *firmitas* (solidity);
- *utilitas* (functional usefulness);
- *venustas (delight)*

The criterion of solidity, which in this case derives from the basic data for the design and implementation of the works, was achieved both at the construction stage and the transformation stages.

Bearing in mind the task of constructing a below-ground connection in a particularly complex context, the second criterion of functional usefulness was particularly exalted, given the significant change that took place over a short period, from a demand for moving freight to and from the Porto Antico to the increased demand for public transport.

The decision to use the existing tunnels as much as possible, suitably modified, enabled them to be brought into step with these changing times and for that reason, to still be functionally useful.

We also find it interesting that ideas were also put forward for making use of the very short stretch of the Delle Grazie Tunnel that goes from the end of the escape route, on the seaward side, to the intersection with the Genoa Metro system.

These works mean that improvements have now been made, in a range of different urban contexts, to below-ground routes no longer in use.

The proof of these improvements can now be seen in the fascinating tourist itineraries that are on offer from the city hosting this conference.

And the possibility, even though it would present some difficulties, of opening the disused section of the Grazie Bassa Tunnel via the vertical shaft at Piazza Corvetto, for shared and regulated public use, could create similar new itineraries that would serve as historic reminders of the engineering history of Genoa.

If that were possible, we could certainly claim to have left evidence behind us that the third requirement, *venustas* (beauty) had been applied.

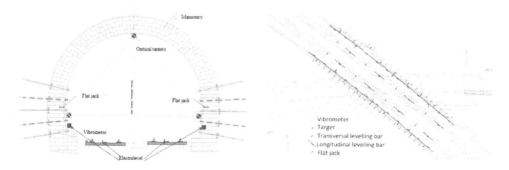

Figure 10.   The monitoring system for the existing Traversata tunnels underpass by the rehabilitation works performed in the Grazia Basse tunnel: cross-section and plan.

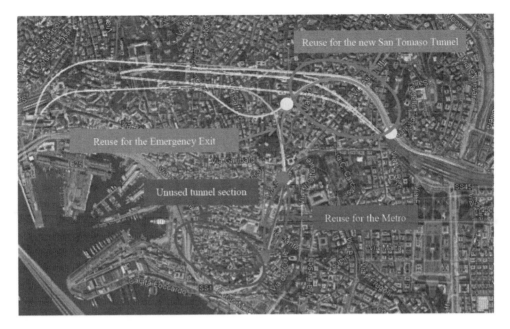

Figure 11. Final plan of Genoa tunnels infrastructure between Brignole and Principe stations.

REFERENCES

Pigorini, A., Pedone, E. & Mohamed Dahir, H. 2010. Aspetti progettuali delle opere in sotterraneo inserite nel potenziamento infrastrutturale del nodo ferroviario di Genova. *Gallerie e grandi opere sotterranee 93.* Bologna. Patron editore.

Pedone, E., Mohamed Dahir, H. & Beltrandi, N. 2014. Risultanze in fase realizzativa dello scavo delle opere in sotterraneo inserite nel Potenziamento Infrastrutturale del Nodo Ferroviario di Genova. *Gallerie e grandi opere sotterranee 109.* Bologna. Patron editore.

Anonymous 2015. 1942-Il-tragico-massacro-della-galleria-delle-Grazie. *www.genovaquotidiana.com*

Pigorini, A. et al. 2016. Upgrading of city of Genoa railway junction: under-crossing with shallow distance of existing tunnels in urban environment. *Tunnel congress ITA-AITES world 2016 (WTC 2016) proceedings:* 527–537. Englewood. Society for Mining, Metallurgy and Exploration Inc

Gaggino, E. (n.d.). Galleria delle Grazie. www.ferrovieinrete.com

Tunnels and Underground Cities: Engineering and Innovation meet Archaeology,
Architecture and Art, Volume 10: Strategic use of underground
space for resilient cities – Peila, Viggiani & Celestino (Eds)
© 2020 Taylor & Francis Group, London, ISBN 978-0-367-46878-1

The Genoa Bypass project: The new highway system to overcome the congestion in the Genoa area. The geological and engineering challenges

A. Selleri
Autostrade per l'IItalia S.p.A., Rome, Italy

ABSTRACT: The Genoa Bypass project, known as the "Gronda di Genova", has been promoted by the concessionaire *Autostrade per l'Italia S.p.A.* to overcome the congestion of the highway system in the Genoa area.

After a complex process of public debate and interaction with the territory, the design was developed to a further stage and in the 2016 it was submitted to the Ministry of Transportation for the final approval.

The new motorway network develops almost entirely underground, through 25 new tunnels having a total length of 50 km.

From the geological standpoint, the Genoa Bypass route crosses a region of great structural complexity, where the Alpine orogenic domain is juxtaposed to the Apennine domain.

The construction works will be divided into five functional lots and will have an overall duration of 10 years; the tunnels excavation will proceed simultaneously on multiple fronts, starting from the construction sites strategically located in the project area.

1 PROJECT OVERVIEW

The Genoa Bypass project, known as the "Gronda di Genova", is currently at an advanced design stage, after the conclusion of the first public Italian project debate and interactions with the territory, the local authorities and Genoa's citizens.

The investment promoted by the highways concessionaire *Autostrade per l'Italia S.p.A.* aims to improve the efficiency of the existing highways, to separate through traffic from local traffic, to support the economic growth and to improve the road safety by providing an alternative route to the only available one along the Ligurian coast.

The new highway system, with two lanes bi-directional carriageways, will double the existing A10 Genoa-Ventimiglia route in the Genoa urban section and will enhance the sections of the A7 Milan-Genoa and of the A12 Genoa-Rome routes, between the East Genoa, West Genoa and Genoa Bolzaneto junctions (see Figure 1).

Between 2009 and 2015, all the required approvals were granted to the Genoa Bypass project, which is compliant with all the requirements of the Environmental Impact Assessment and of the multi-authorities Planning Conference; in 2016 the project has been submitted for approval to the Ministry of Infrastructure.

The new road system runs almost entirely underground, with 25 new tunnels having a total length of 50 km and cross sections of up to 500 square meters in the diversion and connection caverns. Works for the tunnel construction will account for 9 million cubic meters of excavated rock, which will be reused within the framework of the project itself. The geological study envisages that the mineral composition of about 50% of the rock mass along the route could include asbestos fibres.

Figure 1. General view of the Genoa Bypass route.

In addition to the underground works, civil works involve the construction of one new cable-stayed bridge, 10 new road bridges and the improvement and deck widening of 10 existing bridges.

2 GEOLOGICAL SETTINGS

From the geological standpoint, the area involved in the construction of the Genoa Bypass is part of a region of great structural complexity, where the Alpine orogenic domain is juxtaposed to the Apennine domain (see Figure 2).

From West to East, three sectors with specific geological and structural features are encountered (see Figure 3):

– the Voltri Group;
– the Sestri-Voltaggio zone;
– the Apennine Flysch domain, consisting of different tectonic and tectonic-metamorphic units stacked with European vergence (approximately E-W in the current state).

Late and post-orogenic deposits are also present; they could be ascribed to the Tertiary Ligurian-Piedmontese basin and to marine and continental deposits of different ages, which have partially covered the bedrock units and filled the paleo-valleys.

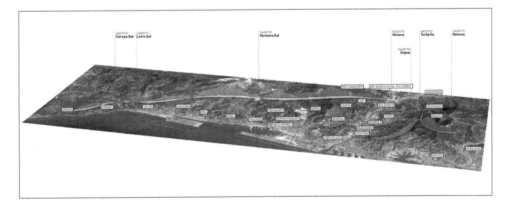

Figure 2. Overall view of the longitudinal profile of the tunnels.

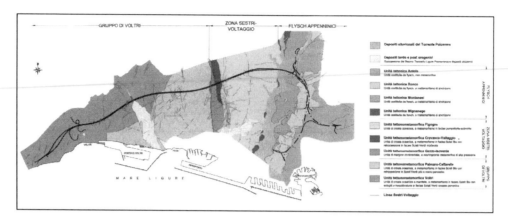

Figure 3.   Geological map of the Genoa Bypass area.

The Voltri Group and the Sestri-Voltaggio Zone are separated by a N-S structural lineament, the so called "Sestri-Voltaggio Line", which is identified as the physical limit between the Alps and Apennines.

In the geological literature, such tectonic lineament has been interpreted in different ways: as a transform fault, as a thrust or as an extensional fault. Currently it is described as part of a complex system.

The geological sequences belonging to the Voltri Group are referable to the Piedmontese-Ligurian domain, i.e. an ocean basin, consisting of native intrusive and extrusive rocks and sediments of different types, deposited over the basaltic lava; in the literature, many tectonic-metamorphic units have been distinguished inside this unit, traditionally referred to two main groups:

– units consisting mainly of metagabbro and serpentine rocks (Beigua Unit, Ponzema Unit, S.Luca Colma Unit);
– units consisting mainly of calcareous schists and prasinites (metabasite), but also of subordinated mantle origin rocks such as serpentinite and peridotite (Alpicella Unit, Ortiglieto Unit, Palmaro-Caffarella Unit, Voltri-Rossiglione Unit).

The metamorphism grade reached the blue shale facies (up to eclogites for the Voltri unit), subsequently downgrading to the green shale one.

The Sestri-Voltaggio zone affects the mountain ridge that forms the upper part of the right hydrographic side of the Polcevera valley and includes tectonic-metamorphic units, highly different for lithology and degree of metamorphism:

– Monte Gazzo tectonic unit, consisting of Triassic dolomite and limestone;
– Cravasco-Voltaggio tectonic unit, made up of serpentinite, meta-basalt, phyllitic limestone and shale;
– Monte Figogna tectonic unit, consisting of serpentinite, metabasite and shales, sometimes with thin carbonate and siliciclastic levels.

The metamorphism grade progressively decreases from the Triassic carbonate units to the ophiolitic series of the last two tectonic metamorphic units.

The Flysch Apennine Domain affects the Polcevera river valley and includes a series of tectonic and tectonic metamorphic units, that are very homogeneous from the lithological standpoint, with a metamorphism grade progressively decreasing proceeding from West to East; the Polcevera river left bank units may be considered as non-metamorphic. These units are stacked with East to West vergence and form bands elongated approximately in North to South direction along the Polcevera valley.

3 MANAGEMENT OF THE EXCAVATION IN ROCK MASS CONTAINING ASBESTOS MINERALS

Due to their genesis, the mineral composition of the rock mass in the west Polcevera mountains shows the potential presence of asbestos (see Figure 4). The Genova Bypass project requires the management of more than 5 million cubic meter of hazardous material due to asbestos mineral presence, coming from the tunnels excavation in that area.

Nowadays, the technology allows suitable management of the excavation of hazardous rocks and therefore this critical issue can be overcome.

In order to operate in a healthy and safe environment during all the tunnel construction stages, from the excavation in the tunnels to the final excavated material disposal, two Slurry-shield TBMs will be used in those west side tunnels of the Genoa Bypass.

The material coming from the TBMs excavation chamber will be directed, under hermetic conditions through pipelines installed along the tunnels under construction, to the Bolzaneto jobsite, where it will be temporarily stored in special silos, thus preventing the contamination of any external environment. At the Bolzaneto jobsite, after being dried via mechanical industrial process, the excavated material will be sampled, analysed and classified, according to the asbestos materials management regulation requirements, to define its final disposal and reuse (see Figure 5).

According to the classification results, three alternative destinations have been identified (see Figure 6):

– sea reclamation area (green code, asbestos content lower than 1 g/kg): the material suitable to be reused as infill for the Genoa airport runway expansion will be pumped through a sealed slurry pipeline to the inshore reclamation area in the Genoa harbour alongside the airport;
– tunnel invert (yellow code, asbestos content higher than 1 g/kg): the material, whose geotechnical characteristics are compliant with the design technical specification to be reused onsite, will be used as tunnel invert filling material, suitably protected by a layer of non-contaminated material and the road paving;
– landfill (red code, asbestos content higher than 1 g/kg): the non-reusable material, whose geotechnical properties do not comply with the design technical specification, will be sent to landfill for authorized disposal.

In order to ensure an environmental safe transportation of the reusable material from the Bolzaneto jobsite to the reclamation area, a hydraulic system running along the Polcevera

Figure 4. Geological-structural map of the Genoa Bypass area, highlighting the potential susceptibility to the presence of asbestos (green arrow).

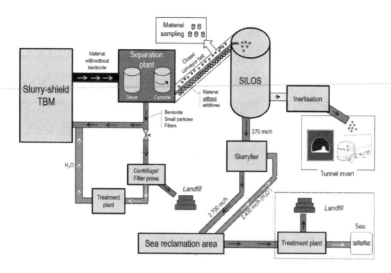

Figure 5. Scheme of the excavated material management.

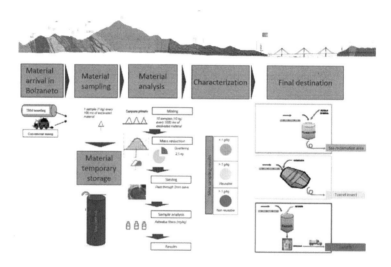

Figure 6. Excavated material sampling, characterization and final destination.

river is foreseen. The dried excavated material will be "slurryfied" by adding sea water and will be pumped through a 10 km long sealed pipeline, to be disposed into the reclaimed basin in front of the airport.

At the works' completion, the slurry pipeline will be dismantled and the jobsites will be reconditioned.

4 SESTRI-VOLTAGGIO ZONE CROSSING

The Monterosso tunnels pass through the challenging Sestri-Voltaggio zone described above.

The section in the red box (see Figure 7) lies inside the tectonic-metamorphic units that form part of the boundary zone between the orogenic Alpine and Apennine domains:

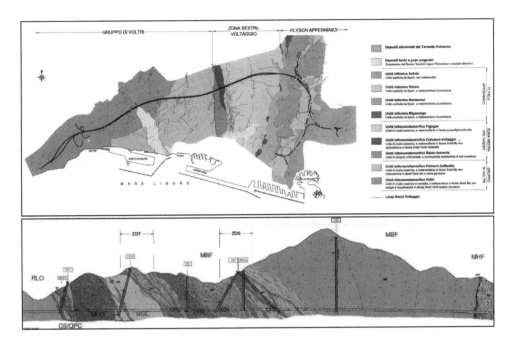

LEGEND

Palmaro-Caffarella Unit: RLO: serpentinite; CS: schists; QPC: quartz-schists.
Gazzo-Isoverde Unit: MDG: M. Gazzo dolomitic marble; SGL: Gallaneto-Lencisa series.
Cravasco-Voltaggio Unit: SPV: Case Bardane serpentinite; LRV: Larvego shale.
Figogna Unit: AGI: Costagiutta shales; SPF: Bric dei Corvi serpentinite; MBF: M.Figogna meta-basalt; MHF: siliceous shales; ERZ: Erzelli meta-limestones.
Tectonized zones: ZDT: Timone deformation zone; ZDS: Scarpino deformation zone

Figure 7.   Geological map and longitudinal profile of the Sestri-Voltaggio zone.

- the Voltri Group on the west, comprising mantle and oceanic crust units and marine meta-sediments, referring to the Ligurian-Piedmont ocean basin;
- the Sestri-Voltaggio Zone on the east, consisting of different origin tectonic flakes, comprising sedimentary deposits and oceanic basement units.

Figures 8, 9 and 10 show pictures taken during the preliminary surveys of the area.

Highly deformed and tectonized zones are present along the tunnels route; the investigations performed so far allowed to identify all the critical conditions that could be encountered during the tunnels excavation, to define the procedures and measures to be implemented in order to face any of these occurrences and to estimate the cost with a high degree of reliability.

Despite the big amount of geological, hydrogeological and geotechnical investigation performed, the Sestri-Voltaggio zone crossing still represents an element of uncertainty, in terms of:

- exact location of geological limits;
- hydraulic and geomechanical characteristics of the rock masses;
- hydrostatic load in high overburden conditions.

A specific complementary investigation campaign has been planned to obtain further information directly at tunnels' level, to improve the knowledge about the geotechnical conditions.

The main aims are the following:

- to obtain a detailed definition of the geomechanical conditions along the tunnel section of particular geo-structural complexity, by extremely long core-recovery bore-holes with BHTV (exact location of the lithological contacts, definition of the fracturing degree of the

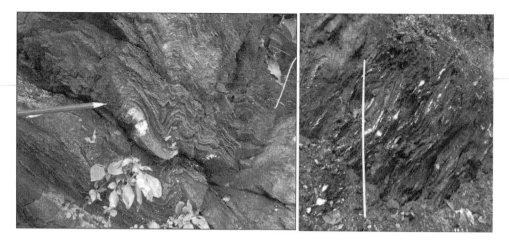

Figure 8.   Folds in the foliation of calc-schists and outcrop of Murta clay shales.

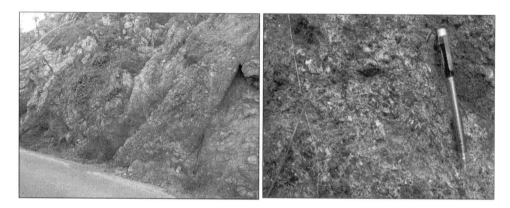

Figure 9.   Outcrop of brecciated serpentinite and detail.

Figure 10.   Outcrops of serpentine cataclasite, incorporating a big lithon and milonized zones on the right.

rock masses, confirmation of the asbestos fibres presence, definition of hydrogeological settings, etc),
–   to verify the design assumptions relating to hydrostatic load, particularly in high overburden condition, by the installation of a particular type of piezometers, designed to ensure the effective partitioning of the aquifers,

- to determine the rock masses permeability and to confirm their capacity of being subject to permeation grouting treatment and to draining patterns, as per the design assumptions, by special in situ tests,
- to improve the quality and reliability of the strength, deformation and in situ stress rock masses parameters for the geotechnical calculation models, by dilatometer tests, hydro-jacking tests, seismic tomography in reflection and refraction, D-H and C-H geophysical tests.

In addition to the in situ investigations, complementary geotechnical and geomechanical laboratory tests have been planned.

The campaign is currently ongoing, so that in the next Detail Design stage of the project, the new information will allow to confirm the design assumptions and to fine-tune the measures adopted to overpass the identified critical conditions.

5 DIVERSION AND JUNCTION CAVERNS

Another technical challenge of the tunnels' construction is the excavation of the diversion and connection caverns that allow the connection between different sections of the new highway system and the existing road network (see Figure 11).

Figure 11.   Caverns' location along the Genoa Bypass route.

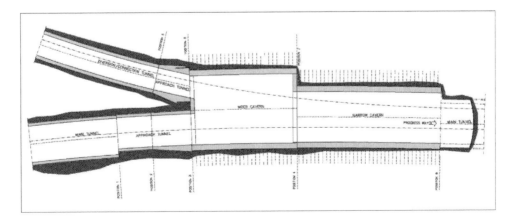

Figure 12.   Cavern typical layout – plan view.

Stage	Working sequence	Cross section

1   Excavation and lining of the main tunnel, up to the cavern approach section
Excavation and lining of the approach tunnel section

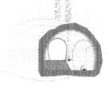

2   Excavation of the crown tunnel
Rock mass improvement treatment

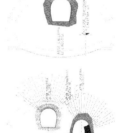

3   Excavation of the right side tunnel
Rock mass improvement treatment
Cast of the right cavern abutment (reinforced concrete)

4   Excavation and lining of the diversion/connection tunnel, up to the cavern approach section
Excavation and lining of the approach tunnel section

5   Excavation of the left side tunnel
Rock mass improvement treatment
Cast of the left cavern abutment (reinforced concrete)

6   Partial refilling of the side tunnel
Cavern crown excavation
Cast of the cavern crown lining (reinforced concrete)

7   Prosecution of the excavation of the right side tunnel in the cavern narrow section
Rock mass improvement treatment
Cast of the right cavern abutment (reinforced concrete)
Excavation and cast of the invert of the wider cavern section (reinforced concrete)

| 8 | Prosecution of the excavation of the left side tunnel in the cavern narrow section
Rock mass improvement treatment
Cast of the left cavern abutment (reinforced concrete)
Completion of the invert of the wider cavern section | |

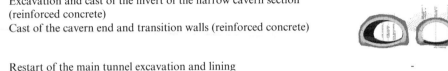

9   Partial refilling of the side tunnels
        Crown excavation in the narrow cavern section
        Cast of the cavern crown lining (reinforced concrete)

10   Excavation and cast of the invert of the narrow cavern section
        (reinforced concrete)
        Cast of the cavern end and transition walls (reinforced concrete)

11   Restart of the main tunnel excavation and lining

Figure 13. Cavern typical construction sequence.

The highway tunnel progressively enlarges to accommodate the diversion and connection ramps aside the two lanes carriageway, thus the cavern cross sections area gradually increases up to 500 square meters (see Figure 12).

The cavern excavation sequence is extremely complex, to guarantee safe conditions during all phases of the construction. It entails different excavation stages and requires performing rock mass improvement treatment, defined according to the local geotechnical conditions.

Figure 13 summarises the typical construction sequence. According to the working plan, the cavern excavation will start from the main tunnel on the wider section of the cavern towards its narrow end.

6 CONSTRUCTION PLAN AND TIME-SCHEDULE

The Genoa Bypass project is divided into five major functional lots and the construction schedule involves, for an overall duration of 10 years, the simultaneous excavation of many tunnels, progressing co-ordinately on multiple fronts at the same time, starting from the job-sites strategically located in the project area (see Figure 14):

– lot 1 (East and West Polcevera river) includes the most critical works in terms of environmental and planning issues (e.g. preparation of jobsites and base camp; excavated material management system, including the installation of pipelines along the Polcevera river and the sea reclamation area filling; preliminary works in Bolzaneto, Torbella and East Genoa areas; all civil works in Voltri and Varenna areas);
– lot 2 (East Polcevera river) includes the first series of conventional mining tunnels and the bridges in Bolzaneto area;
– lot 3 (East Polcevera river) includes the second series of conventional mining tunnels and the bridges in Torbella area;
– lot 4 (East Polcevera river) includes the third series of conventional mining tunnels and the bridges in East Genoa and West Genoa areas;
– lot 5 (West Polcevera river) involves the use of two TBMs to excavate three twin tube tunnels, thus doubling the existing A10 highway.

Lot 1 will be performed during the very first project stages and will start at the first semester, while the other lots will be awarded afterwards and will be constructed independently, in mutual coordination.

Figure 14. Major functional lots and construction sites of the Genoa Bypass.

From the third semester, the construction of lots 2, 3 and 4 in the East Polcevera river area will start, for the construction of highways A12 and A26 road links and the Genoa harbour junction. Construction works of lot 2 will start from the Bolzaneto jobsite and progress southward. Meanwhile, tunnel excavation works of lot 3 will begin at the Torbella area and will progress both northwards, to join lot 2, and also southwards. Here they will meet lot 4, progressing from the tunnel portals of East Genoa Est and West Genoa and from the Morandi Bridge portal.

Also the construction of lot 5 will commence in the third semester. The first TBM will be launched from Bolzaneto on the fourth semester, while the second TBM will start on the fifth. TBMs breakthrough in Voltri is expected during the thirteenth and fourteen semesters. Here, both TBMs will be partially dismantled, reassembled and towed further on over the Leiro and Cerusa bridges. They will then resume the excavation and on the fifteenth semester are expected to reach Vesima, where they will be finally decommissioned.

In the sixteenth semester, the construction of all the major civil works is expected to be completed. After the completion of all the facilities, installation and finishing works, the Genoa Bypass will be opened to traffic in the twentieth semester.

7 CONCLUSION

The present paper focuses on the main geological and engineering challenges of the Genoa Bypass project, the so called "Gronda di Genova", promoted by the concessionaire *Autostrade per l'Italia S.p.A.*

After an interactive process, involving the territory, the local authorities and Genoa's citizens, the project has been developed to a thorough design stage and, finally, in the 2016 was submitted to the Ministry of Transportation for the final approval.

The new motorway route develops almost entirely underground, crossing a region of great geological complexity.

Advanced technical solutions have been adopted to manage the excavation of hazardous rocks, with possible presence of asbestos fibres.

Highly deformed and tectonized zones are present along the tunnels route, particularly in the Sestri-Voltaggio Zone. An additional investigation campaign is currently ongoing to obtain more detailed geological information, directly at tunnels' level, to confirm the design assumptions and to fine-tune the measures to overpass the identified critical conditions.

The construction works have an overall duration of 10 years and will be divided into five functional lots. The tunnels excavation will proceed simultaneously on multiple fronts, starting from the construction sites strategically located in the project area.

After the completion of all the civil works and of the facilities, installation and finishing works, the Genoa Bypass will be opened to traffic, allowing overcoming the present congestion of the highway system in the Genoa area.

REFERENCES

Capponi G., Crispini L. (2007) - Note illustrative della Carta Geologica d'Italia alla scala 1:50.000 - Foglio 213–230 "Genova". APAT Servizio Geologico d'Italia, Roma, 178 pp.

Carta Geologica d'Italia alla scala 1:0.000 – Progetto CARG - Foglio "Genova e relative Note Illustrative".

Crispini L. (1996) - "Evoluzione Strutturale dei Metasedimenti del Gruppo di Voltri e della Zona Sestri-Voltaggio: implicazioni nell'evoluzione Tettonica e Geodinamica Alpina." PhD thesis, published by ERSU, Genova Univ., 129 pp.+ 126 fig.

Federico L., Capponi G., Crispini L., Scambelluri M. & Villa (2005) - 39Ar-40Ar dating of high-pressure rocks from the Ligurian Alps: evidence for a continuous subduction - exhumation cycle. Earth Plan. Sci. Lett., 240, 668–680.

Giacomini F., Boerio V., Polattini S., Tiepolo M., Tribuzio R. & Zanetti A. (2010) - Evaluating asbestos fibre concentration in metaophiolites: a case study from the Voltri Massif and Sestri–Voltaggio Zone (Liguria, NW Italy). Environmental Earth Sciences, v. 61, n. 8, pp. 1621–1639.

Marini M. (1998) - Carta geologica della Val Polcevera e zone limitrofe (Appennino Settentrionale) alla scala 1:25000. Note illustrative. Atti Tic. Sc. Terra, 40, 33–64.

Perello P., Venturini G., Dematteis A., Bianchi G.W., Delle Piane L. & Damiano A. (2005) - Determination of reliability in geological forecasts for linear underground structures: the method of the R-Index. Atti del simposio internazionale "Geoline 2005", Lyon (F), 23–25 maggio.

Testa C., Carlucci P. L., Pelizza S., Peila D., Kalamaras G. & Alessio C. (2008) – Tunnel design in asbestos-bearing rocks and design of an underground cavern for storing the contaminated muck material. The case of the Cesana tunnel. Congrès International de Monaco. Octobre 2008.

*Tunnels and Underground Cities: Engineering and Innovation meet Archaeology,
Architecture and Art, Volume 10: Strategic use of underground
space for resilient cities – Peila, Viggiani & Celestino (Eds)
© 2020 Taylor & Francis Group, London, ISBN 978-0-367-46878-1*

Istanbul Airport Express Line: Design and built of the new strategic mass transit underground connection system

E.G. Trussoni & M. Palomba
Geodata Engineering SpA, Turin, Italy

A.G. Durgut
Geodata Engineering SpA, Istanbul, Turkey

G. Astore
Geodata Engineering SpA, Turin, Italy

ABSTRACT: The growth of the Turkish infrastructures continues its ascent. In this scenario the Gayrettepe–Istanbul New Airport line covers the role of main strategic public transportation link under construction. The main purpose of this line, located on the European side of Istanbul, is to provide direct and quick connection from the city center to the largest hub airport in the world under construction, with 200 million annual passengers. The 38 km long twin-tube line, other than service/emergency stations and crossover structures, includes underground stations located in urban, suburban context and terminals area. Among several complexities which have been successfully solved during design and construction stages, the attention is mainly focused on Kağıthane station, a cut & cover box excavated in sensitive geomorphological area in heterogeneous grounds and on the rectification design for M10 crossover cavern, developed to minimize the impacts on the construction schedule of the new airport.

1 GAYRETTEPE–ISTANBUL NEW AIRPORT EXPRESS METRO LINE TYPE AREA

1.1 *The environmental and urban context*

Istanbul is a megalopolis with over 15-million inhabitants, characterized by a mixture of its historic heritage, natural beauty and uncontrolled urbanization. Mobility is the biggest problem of the city, also due to its unique location, divided in two parts by the Bosporus Strait, because of continuous exponential population growth and for lack of planning over the past decades.

Along the past 12 years, Istanbul Metropolitan Municipality (IBB) and General Directorate of the Infrastructural Investment (AYGM) funded a large investment plan to address the increasing and urgent demand for improved connectivity: to build an integrated mass transport network, covering the metropolitan area, extending for approximately 500 km to bring into service in 2023 (100th celebration of the Turkish Republic founding).

Within this ambitious investment plan, Geodata Engineering is involved, as Designer, in the implementation of new Express Metro Line (M11) Gayrettepe–Istanbul New Airport, on the European side of the city, currently under construction by Şenbay-Kolin- Contractors JV.

The activities carried out by the Designer have included the revision of the tender design in order to solve several functional and alignment changes required by AYGM, the development of the final and detail design of the overall civil and E&M works, the technical assistance to the Con-tractors during construction based on the real conditions encountered on site.

The Express metro line runs from south to north: starting from the center of Istanbul, Gayrettepe district, crossing the municipalities of Kemerburgaz and Göktürk and arrive up to

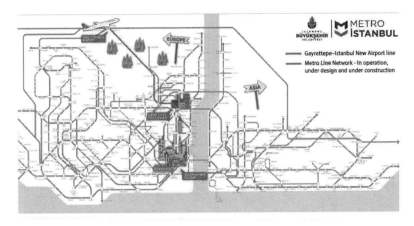

Figure 1. Metro network – Final configuration when all lines will be constructed and in operation.

the new Airport Terminals. Among these urban areas, a major project of a hospital and university center is being developed in the close surrounding of the City and as well a commercial complex is under design in İhsaniye municipality. The remaining portion of the territory is occupied by green and wooded areas.

1.2 General Overview of the project

The Istanbul Airport Express Line is approx. 38.0 km long, twin tube excavated with TBM, located entirely underground at an average depth of about 40m. It includes 9 under-ground stations, most of them to be built in C&C favored by the availability of large sur-face areas, while the only one built in the cavern is Gayrettepe station, due to the high urban density at ground level.

According to the Express Metro operation requirements, along the line have been fore-seen 09 crossovers structures (equipped with installations enabling metro trains to be guided from one track to the other – built in cavern and in C&C), 10 Service Stations (equipped with installations to manage line power supply, train piston effect relief and emergency escape ways for passengers – built through a shaft/caverns complex system), 4 shafts fully dedicated to the emergency (in consideration of the big distance between the station to permit higher level of safety for people, in comply with the international standard requirements).

The system adopted is GoA4 fully automatic Train Control System and all station platforms (180m) are provided with Platform Screen Doors.

The strategic placement of this line, inside the Istanbul Transportation planning, is demonstrated by the number of passengers estimated: approx. 1.100.000 Pax/day. Traveler will be able to reach the Airport Terminal within 30 min with and headway of 3.00 mins during peak hours.

1.3 Challenge of the project

Main goal to achieve with the Express Line may seem to be just to guarantee a comfort-able and fast public rail connection from Istanbul City to the new International Airport. In the main picture of all Istanbul, this line will play the strategic role to guarantee a stronger interchange system inside the actual a future public transportation network:

– Havalimani 1: interchange station with High Speed Train (Edirne-Ankara).
– Kağithane: interchange station with Istinye-ITU-Kağithane Metro Line (under design) and M7 viaduct line (under construction).
– Gayrettepe: interchange station with M2 line (in operation), BRT Metrobus on grade (in operation) and 3 Storey Great Istanbul Tunnel, a single tunnel which will include one layer for the metro line and 2 layers for highway roads (under design). This interchange will

Figure 2. (M11) Gayrettepe–Istanbul New Airport, Express Metro Line.

create the second underground connection between the suburban district in Asian and European sides, passing through the Bosphorus and City center.

1.4 *Geological context*

The tunnel alignment runs through extremely heterogonous geotechnical conditions and with overburden ranging from 15 to 80m, which strictly depends on particular orographic conditions. From Gayrettepe to Kağıthane the tunnels lay in the so-called Trakya Formation (Ct), in a good rock mass mainly composed by sandstone and shales.

In Kağıthane area, 200m stretch of sand and clay are intercepted within the alluvial material (Qks). Then the excavation predominantly proceeds in the Trakya formation, but also intercepting local silty and sandy-clay layers close to the Hasdal station and crossover cavern M3, within the so-called Danişmen formation (Td), made by medium to hard clay and clay-stone.

From Kemerbugaz station the Danişmen formation extends up to Havalimani-1 station, with still residual presence of Trakya formation and minor stretches of tunnels excavated in Kiraç Formation, made by hard clay, claystone and siltstone, with risk of local lenses of coal to be intercepted.

The excavation proceeds up to Havalimani-3 in claystone, marns and limestone, within the Ceylan Formation.

1.5 *Running tunnels*

The twin tube single track option has been foreseen for the running tunnels throughout the entire length of the express line. The twin tubes are predominantly (67.5km) excavated by 10 EPB-TBMs provided by Herrenknecht, Terratec and NHI, all assembled in C&C stations.

Four TBMs have been launched from Ihsaniye station, four will be launched from Kemerburgaz station, and other two will be assembled in Hasdal station and they will mine towards Kağıthane station. The TBMs will pass through the already excavated crossover caverns and service emergency stations, were maintenance of machines will be also carried out. All the TBMs will emerge to ground surface through C&C stations and dedicate retrieval shafts.

The balance of the running tunnels from Kağıthane to Gayrettepe station (2.7km) are bored by NATM, due to the very dense urban area which does not allow installation of jobsite logistics for launching of TBMs.

The TBM tunnel lining consists of a universal ring made by reinforced concrete trapezoidal segments (5+1), with an outer diameter of 6.3m, an inner diameter of 5.7m and a width of 1.5m. For the NATM running tunnels (including crossover caverns, service stations and cross passages), the permanent lining is made by a cast-in-place reinforced concrete and it is sealed against potential water inflows with the so-called "full-round" water-proofing system, which consists of a drainage and protection geotextile layer and a sealing PVC membrane to be placed between the primary support and the final lining, along the whole tunnel perimeter.

Figure 3.   Excavated running tunnel and TBM breakthrough in crossover cavern.

1.6   Underground Stations

The C&C stations are typical rectangular boxes, which are constructed with bottom-up methodology. Under this method, after the construction of the temporary retaining structure, the open-cut excavation is conducted up to the bottom level with the simultaneous installation of the anchoring/strutting system.

The permanent structure is independent from the support of the excavation and the construction proceeds from the bottom slab up to the top slab, with the gradual removal of the anchoring/strutting system.

In detail, the definition of the sequence of excavation of Ihsaniye and Kemerburgaz stations, has been the real challenging aspect. As a matter of fact, the overall stability has been checked for the different stages, considering the actual locations of struts which were installed with non-regular and uncommon arrangement to achieve the required spaces for assembly of TBMs (paying attention for lowering and lifting of each single piece of shield and back-up elements), launching structures (thrust frames) and conveyor belts operations.

All the potential interferences with TBM's logistics were successfully solved through a very effective cooperation between Designer, Contractor and TBMs' manufacturer.

1.7   Crossover Caverns

To minimize the disturbance induced by the excavation to the surrounding ground and for construction reasons due to the big size of the excavation (18m span and 12m high for approx. 170m2 cross section area), the construction of the crossover caverns has been carried out with the so-called "sequential excavation method" (SEM) based on NATM.

Due to the heterogeneity on the geological and geotechnical conditions, different excavation sequences based on multi-drift approach have been analyzed through numerical analyses and different types of supports were defined.

The widely applied temporary support for the crossover caverns is made by 30cm thick. shotcrete reinforced with double layer of wire mesh Q221/221, steel profiles HEB160 with spacing ranging from 0.6m to 0.8m to be selected in accordance with the overburden and geological conditions and systematic bolting and drainages.

Additional pre-support measures, such us pipe roof, soil nailing at tunnel face and ground improvement treatments were also implemented in the sectors where particular unfavorable

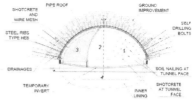

Figure 4. Sequential Excavation Method applied for crossover cavern.

conditions were expected. The excavation of the crossover caverns starts from a dedicate access tunnel which is excavated passing through temporary shafts.

1.8 Service stations

In order to simplify the construction activities, the same layout has been applied for all the service stations. It includes 20m dia. shaft (for access, emergency evacuation and ventilation purpose), from which a reticulum of "service tunnels" is progressively excavated.

The excavation of shafts has been very challenging, mainly due to the considerable depths (50m). Consequently, for practical construction reasons and to minimize the vertical misalignment of piles, a sort of "telescopic-system" has been adopted which consists in providing more intermediate drilling levels.

The excavation in the shallow portion (in soft soil) is conducted within spaced-piles retaining wall with different levels of reinforced concrete rings to be used as a propping system.

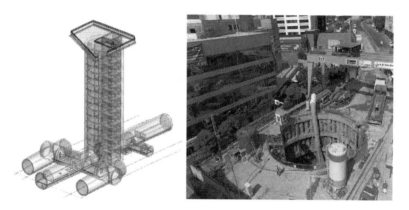

Figure 5. Typical layout of a service station and shaft excavation and support.

The bottom portion of excavation (where generally slightly better conditions were expected) is supported by minipiles, only where required, shotcrete reinforced with wire mesh and systematic bolting and drainages. In the worst geological conditions, ground improvement treatments (such as jet grouting columns) have been considered essential to deal with the excavation, preserving the integrity of the soil/rock pillars between the different tunnels as well as the stability of the tunnel junctions.

2 KHAGITHANE STATION

2.1 Station location and geological context

Kağıthane is a typical a Cut & Cover station located between Gayrettepe and Kemerburgaz approx. at Ch. 3+800. The location of this station strictly depends on complex geomorphological conditions of the area and on existing utilities, which were not possible to di-vert. In particular, the vicinity to the Kağithane river and the presence of an existing 1800mm dia. wastewater collector, implied the necessity to move the C&C box towards SE direction, beneath the existing slope.

Due to this particular location, the station is excavated in very heterogonous condition. The SE side (hill portion) is excavated in the so-called Trakya Formation, constituted by sandstone, siltstone and layers of claystone. This formation is subdivided in unit Ct1 (weathered rock mass) and Ct2 (fair rock mass). The NW side, close to the river bank, is excavated in extremely poor alluvial formation com-posed by soft clay, silty-clay, sand and gravel, subdivided in Qal1 and Qal2, depending on the degree of weakness of soils deposit.

from which a reticulum of "service tunnels" is progressively excavated.

2.2 Arc-shape protective structure

Based on the available geological investigations carried out during the Tender Design stage, potential active landslides were preliminary identified in the Trakya Formation, on the hill side. Consequently, during the Detailed Design stage, several back-analyses were carried out in order to confirm the expected location of the slip surfaces, evaluating also the effects induced by the station open-cut excavation on the overall stability of the existing slope. After having identified a deeper slip surfaces at the interface between weak (Ct2) and more competent rock (Ct1), the achieved results implied the necessity to design a massive protection measure in order to nullify the risk of instabilities and preserve the existing buildings in the area.

Therefore, the idea was to designs a massive arc-shape piles- retaining wall which had to be constructed prior to start with the station excavation. The retaining wall is made of 1000 mm dia. piles, 1200 mm spaced, with length variable from 11.0m to 35.0m, in order to achieve sufficient socket length in sound rock and also to intercept the deeper slip sur-faces. Due to the considerable height of the excavation, four levels of anchors, with a bond length 8.0m were also proposed to ensure intermediate supports during the subsequent excavation stages, along with systematic drainage pipes to relieve water pressure be-hind the retaining structure. Additional surface drainage measures, based on environmental engineering concepts, were also proposed to properly collect the waters and avoid risk of erosion and shallow instabilities.

As mentioned before, the main scope of the arc-shape retaining wall was to stabilize the existing slope and preserve the existing structure from potential damages induced by station excavation. In fact, the arc-shape configuration allows to effectively conveying the stresses outside from the area of the station, avoiding unfavorable overloads to station retaining structures (D-walls and piles wall). The second purpose, strictly related to the construction site logistic, was to provide flat working planes at different level for safely operating the big drilling machines.

The excavation of the Kağithane Station is conducted within diaphragm walls in soil (using special stop-ends elements to ensure enough grip along the construction joints of adjacent

Figure 6. Unfavorable Location of Kağithane Station and longitudinal geological profile.

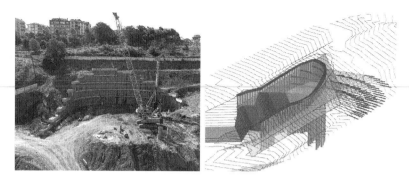

Figure 7. Arc-shape anchored retaining wall and indication of slip surfaces.

panels), while secant piles have been foreseen in rock portion, in order to minimize risks of water inflow, accepting only minor leakage than can be easily managed during the excavation. The retaining walls are supported by temporary steel struts to be removed once the permanent inner structures are constructed. The risk of differential settlements which could be possibly occurred due to the big difference on the bearing capacity of the sub-base soil (rock and soil), has been managed with the construction of piles foundation in the area where the bottom slab is constructed on the alluvial material.

3 M10 CROSSOVER REMEDIAL WORKS

3.1 *Station location and geological context*

The Crossover structure M10 is located between pk 35+668 and pk 35+788, within the new airport area. The excavation of the crossover cavern M10 should have started from the shaft C06 (passing through a temporary access tunnel), to be used for lowering of machineries as well as for ventilation purpose. Therefore, the excavation of the access tunnel and crossover cavern was initially planned to carry out with SEM: top-heading and bench with ground improvement treatments.

In June 2018, thanks to the timely implementation of the monitoring system, some unexpected deformations were registered during the excavation of the access tunnel used to reach the top heading portion of the M10 crossover cavern.

These deformations, which partially affected both shaft and access tunnel, most probably depended on some unpredictable geological conditions: a thicker layer of very weak claystone material (within the so-called Danişmen Formation) was intercepted during the opening of the access tunnel from the temporary shaft C06. In addition, also some ground improvement treatments were found not so effective as it was initially planned.

From above, some cracks occurred in the shotcrete at the junction between the temporary shaft and the access tunnel, therefore the start of cavern excavation (after temporary urgent remedial measures in the most sensitive areas) was immediately stopped to avoid potential collapse or sinking, which could have drastically affected the new airport construction activities. Independently, due to the very tight construction schedule, the Contractor started to excavate an open cut (with very steep slopes) trying to reach as fast as possible the top heading level of the crossover cavern.

Unfortunately, further deformations were registered, and big cracks appeared on the piles of the shaft which were exposed due to the open excavation, with high risk of imminent collapse. GDE immediately suggests the Con-tractor JV to backfill the whole excavation to nullify the risk on instabilities.

Based on above, a new technical proposal had to be defined, since it was agreed that, with the available technology at site, no further underground excavation was considered safe

Figure 8.   Open cut excavation up to reach TH of cavern (piles of shat were dangerously exposed).

enough in a so complicated condition with de-tensioned material due to excavation and back-filling. The remedial option was to try to "convert" the crossover cavern to a standard C&C structure to be excavated within piles-retaining wall, to be drilled all around the cross-over footprint.

3.2   Description of the technical solution

The technical solution was jointly defined by GEODATA Engineering and Contractor JV, with the aims to:

– minimize as much as possible the impact to the airport construction site;
– provide a reliable solution which also ensures high degree of safety for the workmanship;
– provide a cost & time effective solution, in order to partially recover the delay in the construction;
– restore stability conditions in the whole de-tensioned area.

From above, a sort of improvement of the so-called "top-down" construction method of typical C&C structure was proposed. In this method, the underground retaining walls are first installed to work as foundations for the roof slab, which provide support of the excavation. Access openings on the roof slab are provided so that works thereafter could proceed down-wards to the base slab level of the underground structure.

Upon completion of the base slab, the side walls are constructed. The access openings on the roof slab are then sealed and the ground is subsequently backfilled and reinstated.

For the specific case of M10, a massive reinforced concrete canopy, based on the arc structural principle has been found the most suitable solution to cover the huge span of excavation (approx. 19m) without needed for intermediate supports. Once the canopy construction has been completed, the backfilling activities have been immedi-ately carried out to quickly restore the existing ground profile, leaving the site to the Contractor in charge for the construction of the airport. Therefore, the excavation is proceeding underneath the canopy avoiding any kind of further interference since the e access of machineries is achieved through the existing shaft, from which also the muck-ing operations are carried out.

The overall complex is made by:

– 1200mm dia. tangent piles retaining walls (all around the crossover complex) with a length of 21m. For front and back walls "Soft–Eyes" solution has been implemented to allow the penetration through the piles of TBMs which come from İhsaniye station. This unique tech-nology uses cut-able Glass Fiber Reinforced Polymer (GFRP) reinforcement as an advan-tageous replacement for conventional steel rebars.

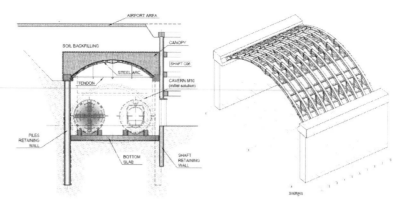

Figure 9. Reinforced concrete canopy, Main elements and steel frame axonometric view.

Figure 10. Canopy construction is ongoing in the new airport site.

- Frame made by steel ribs HEB400 (with same spacing of piles), to be used as a sacrificial formwork for pouring of canopy.
- Tendon steel element (made by coupled UPN 200 profiles, to be installed along the spring line of the arc), which connects both arc foundation beams
- The reinforced concrete canopy structure (with a thickness of 1500mm at center location) which is the key-element that has to withstand to the soil backfilling.

The canopy has been assumed as a permanent structure, therefore the structural capacity has been checked also for long terms loads (water pressure induced by rising of phreatic level and seismic loads). The sealing against potential water infiltrations has been achieved through two components waterproofing coating (polymer modified bitumen rubber) which is quite easy and fast to be applied.

4   CONCLUSIONS

Despite the current problems related to the state of the Turkish economy, which may influence and slow down the ambitious investment program for the new Metro and Light Rail Projects launched in previous years by the government authorities, the ongoing projects in Istanbul are extraordinary experience of design- construction of large-scale infrastructure works in a complex and dense urban context.

Among these, the New Airport express line is undoubtedly one of the most strategically priority and relevant project for the sustainability and improvement of the transport systems of the Megacity.

The present paper deals with some relevant design and construction aspects, emphasizing the applied technical solutions on the challenging underground works to overcome potential risks during the construction stage. Despite the presence of significant obstructions and very demanding contractual terms, the continues support of the Designer and the strong cooperation with the Contractor, are allowing to achieve the works in safety and in compliance with the programmatic objectives of the Government Authority.

ACKNOWLEDGEMENTS

The authors would like to express their deepest gratitude to Şenbay-Kolin Contractors JV for having involved Geodata Engineering in this challenging project, giving the opportunity to deepen the design and construction knowledge on the excavation in heterogeneous ground conditions and for transforming these huge structures into a reality.

Special thanks go to Geodata Engineering Design Teams in Turin Head Office, in Istanbul Branch and India Branch for the effort and precious contribution daily spent.

The authors also wish to acknowledge express Murat Cebeci (PM), Bahriye Yaman (DM), Contractor Design and Construction team for transforming these huge structures in-to a reality and for the fruitful cooperation.

REFERENCES

GEODATA Engineering S.p.A, 2018–2019. Gayrettepe – İstanbul New Airport Metro Line Construction, Procurement, Installation and Commissioning Works of Electromechanical Systems, *Final and Detail design drawings and calculation reports*. Turin and Istanbul.

ŞENBAY-KOLIN CONTRACTORS JV, 2018–2019. Gayrettepe – İstanbul New Airport Metro Line Construction, Procurement, Installation and Commissioning Works of Electromechanical Systems. Istanbul.

Tunnels and Underground Cities: Engineering and Innovation meet Archaeology,
Architecture and Art, Volume 10: Strategic use of underground
space for resilient cities – Peila, Viggiani & Celestino (Eds)
© 2020 Taylor & Francis Group, London, ISBN 978-0-367-46878-1

Comparative analyses of through-running vs. dead-end tunnel to urban transit network efficiency and design of through-running tunnel in the New York Pennsylvania Station

J.C. Venturi
Rethink Studio, New York, USA

E. Chao
Department of System Engineering, University of Pennsylvania, Pennsylvania, USA

C. Nicolescu
Department of Architect, Columbia University, New York, USA

ABSTRACT: Opinions on urban transit network vary the geometric shape, rights of way, tunnel efficiency, and operation strategies, which influence the future scale of urban growth. Major cities cannot function well without tunnels. Studying historical urban transit tunnel operations has revealed a critical topic that has not yet been discussed: how could a through-running tunnel and a regional unified network (RUN) increase both tunnel and transit network efficiency? The article focuses on the comparison of through-running vs. dead-end tunnel operations within the central business district (CBD) and elements of transit network design in the tunnel environment. First, the definition, measures, and decision-making trade-offs are well-interpreted. Second, the time-distance measure shows two independent concepts of static and dynamic transport units on a common variable, allowing tunnel and network designers to precisely define the minimal headway of leading and following trains. The reduced headway decreases the deadhead non-revenue operating time to further increase tunnel efficiency and fleet utilization. Comparative regional rail case studies are presented. Third, schematic station design on platform expansion, track reengineering, and network realignment at the underperforming New York Pennsylvania Station are comprehensively studied. These serial breakdowns enable the establishment of through-running tunnel operation. It is now critical to pay systematic attention to the planning, design, and operation of the interrelation between tunnel efficiency and network throughputs.

1 DEFINITION AND MEASURES OF NETWORK EFFICIENCY

Network efficiency, representing a system's performance as a whole and the cost of transit operations, is a major concern for operators or transit agencies (Musso & Vuchic, 1988; Vuchic & Musso, 1991). There are five outstanding determinants of network efficiency:

Operating flexibility is increased through connections between lines at which Transit Units (TUs) can be switched among lines for scheduled or unforeseen changes in service. More extensive networks and track connections among lines also increases operating flexibility. Turnback tracks (Case I.), usually center dead-end tracks connected with both through tracks, allow short turn operations, i.e. use of intermediate stations as terminals. Storage or pocket tracks can be used for holding reserve trains and storing of disabled trains and maintenance equipment. Frequent crossover (Case II.) tracks between the two main tracks make single track operations easy, facilitating maintenance. Through station (Case III.) connects both end of cities with a much operating pattern. All these types of tracks increase the reliability and efficiency of network operations.

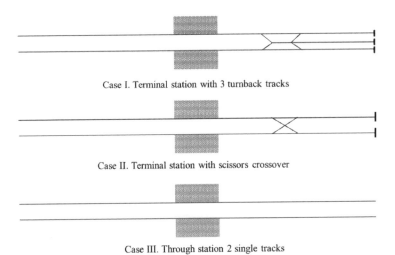

Case I. Terminal station with 3 turnback tracks

Case II. Terminal station with scissors crossover

Case III. Through station 2 single tracks

Figure 1.   Example of flexible operations for track components and alignments.

Continuity and balancing of lines provide direct services among areas with heavy travel demand, which are desirable for both passenger convenience (direct service) and operating efficiency (no transfer delays, superior fleet utilization). Continuity should be embedded into the system's design so that each line has reasonably well-balanced capacity requirements on all of its major sections. The successful operations of any regional unified network need a clear distinction between trunk and branch (feeder) lines, and meticulous network capacity management.

Integration with other modes ensures that a network provides a service in an area greater than intermediate stations. It allows efficient utilization of joint rail and highway corridors and consolidation of intermodal stations, which permit easy interchange of passengers between different modes: bus feeders to rail trunk lines, and park-and-ride possibilities.

Terminals, depots, and yards should be placed at such locations that deadheading, i.e. travel between depots and the lines, is minimized. Because these facilities occupy large areas, they should be located where the cost of land is not high. Both requirements make outlying sites better suited for depot and yard locations than central city areas. Terminals should be placed near major streets to allow access by pedestrians and other vehicle-based modes of transportation.

Cost of the system is, of course, often the most outstanding factor in network design. Investment cost depends mostly on ROW category and alignment: in type (at-grade, elevated, underground tunnel, or other), and stations (size, floors, complexity, equipment, storage facilities, etc.).

The greater the number of passenger trips transit system carries, the better it serves the public and the city. The more person/km a transit network carries, the more economically it operates (as its unit costs decrease) and the more it reduces less desirable private car travel. The ultimate goal in designing most transit systems is to attract as many passengers as possible and increase network efficiency in a tunnel environment. Often, transit network design in a tunnel environment requires finding equilibrium among conflicting requirements, such as overlapping lines for passenger convenience versus operating simplicity of independent lines; or speed, capacity and area coverage versus investment cost. It is crucial that experts with extensive experience in transit systems operations participate in the design of lines and networks, from general line alignments to the details of track layouts, station designs, and operational planning. Figure 2 is an example of finding an optimal station density overlaid with different variables.

Life-cycle costs include the investment and operating costs of the system. It is often a trade-off between the investment and operating cost, and the optimal solution between them should be derived through cost models and analyses. For example, a more expensive alignment allows higher operating speed; or construction of turnback and crossover tracks may not be included in the line design to reduce the investment cost, but the lower operational flexibility may be very costly in system operation, involving extra TU/km of daily travel.

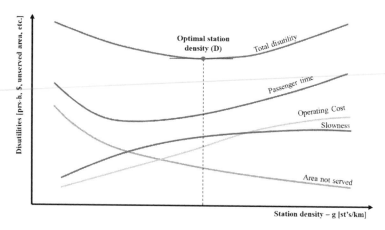

Figure 2.   Example of finding optimal station density under different variables.

Generally, higher investment should provide either lower operating cost or higher service quality or more throughputs and capacity, or a combination of the three, thus attracting more passengers. It is a frequent mistake that investment cost is given excessive relative weight. Although it may be a large sum, investment cost often has only a moderate influence on the total annual system cost, because the long life of facilities makes annual capital cost low. High investment can also be justified by increased passenger attraction, which has a permanent influence on the metropolis economy. This benefit should always be carefully considered (Vuchic, 2007).

2   VISUAL REPRESENTATION OF TRAIN MOVEMENT ON TIME DISTANCE MEASURE

The concept of vehicles on time-distance occupancy address not only the quantity of ground area (i.e. space) that is required for safe vehicle movement or for storage, but the period of time for which the area is occupied as well. An upside of a time-distance measure is the linking of the two independent concepts of static and dynamic transport units (either vehicles or persons) under a common variable. Such a measure allows efficiency to be evaluated in terms of consumed vs. available time-distance in a given timeframe. When city government and transit agency try to quantify the cost of land procurement and consumption ratios for tracks, platforms, yards, facilities, and entire station areas; time-distance measure provides a joint analyses of land consumption by moving and standing passengers and vehicles, which further assist network designers to delineate a dedicated space for rail operations of stations and terminals along the corridor. Time-distance measure offers a visual representation of the overall interrelations of leading and following trains and breakdowns of vehicle occupancies, schedules, and corresponding locations to further inform network density and system throughputs. This is particularly useful in comparing different modes and lines (Benz, 1987; Eric, 1992).

As Figure 3 shown, the time distance representation measures:

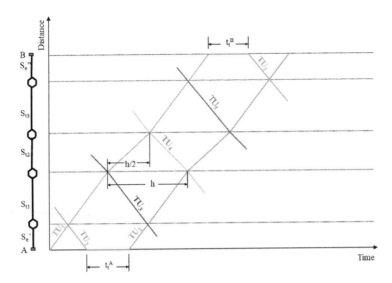

Figure 3. Time-distance diagram for a single line operating with minimum headway of five TUs.

A & B denoted as terminals; $S_{1,2,3,...e}$ denoted as stations; $S_{t1,2,3,...e}$ denoted at standing time; t_t^A & t_t^B denoted as vehicle terminal standing time (also known as deadhead time or non-revenue operating time). As an operator or a transit agency, deadhead time should be minimized.

$$\frac{\Delta S}{\Delta t} = V \tag{1}$$

In Equation 1, V denoted as average vehicle operating speed from station to station;

$$\frac{T}{N} = \frac{60}{f} = h \tag{2}$$

In Equation 2, h denoted as headway measures within a single period, the successive transit units (TUs) at a station (f: frequency), consisting of three groups of elements:

T denoted as time intervals of vehicle (N: number of vehicles) motion, so-called cycle time (i.e. acceleration of leading-TU and deceleration of following-TU), which depend on vehicle dynamic characteristics, operating regime and safety requirements.

Standing time (t_t^A & t_t^B), which consists of door opening, passenger boarding and alighting, and departure preparation (warning, door closing, and departure signaling to the driver).

Signal delays must also be considered.

By measuring these operating elements, a greater understanding of network efficiency can be examined by calculating:

$$\text{Cycle time}: T = \frac{N}{h} = 2T_o + t_t^A \ \& \ t_t^B \tag{3}$$

In Equation 3, T_o is the one-way operating time

$$\text{Route length of round} - \text{trip speed}: l' = \frac{TXV}{120} = \frac{NXhXV}{120} \tag{4}$$

Decreasing standing times and increasing operating and cycle speeds in most situations decrease shown operating costs and may decrease the number of vehicles in service. This operational strategy generates higher fleet utilization rates, which further reduces fleet size and procurement. Ultimately, this measure also results in greater passenger attraction and fare revenues.

3 COMPARATIVE ANALYSES OF THROUGH-RUNNING VS. DEAD-END OPERATIONS ON TIME DISTANCE MEASURE

In rail operation, minimum headway ($h_{s\ min}$), determines the maximum frequency of transit units (TUs) on the line, and is by far the most complex element influencing line capacity. The headway is determined by the relationship between two TUs following each other at a station. The critical situation is when the leading TU is accelerating from the station while the following TU is entering it and braking, so that its stopping shadow area is shown in Figure 4 (blue color), or its movable stopping distance, just touches the path of the rear-end of the leading TU.

The pink line represents the front-end of the leading transit unit (LTU) exiting the station; the orange line represents either the front-end of the LTU or the rear-end of the following transit unit (FTU). When a vehicle travels at a constant speed, its shadow pink line and the position of its front end are parallel. When a vehicle approaches a station and decelerates, its shadow line extends to the far end of the station platform and then becomes a horizontal line. The path of the vehicle's front-end during braking coverages with the shadow line and coincides with it when a vehicle is stopped.

The minimum headway ($h_{s\ min}$) as a function of speed V calculated by:

$$h_{s\ min} = t_s + t_r \tag{5}$$

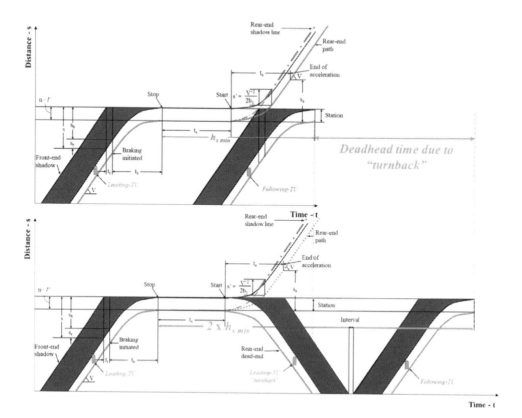

Figure 4.   Time-distance measure on departure/arrival sequence and headway between consecutive TUs comparison at Through-running (up) vs. Dead-end (down) station.

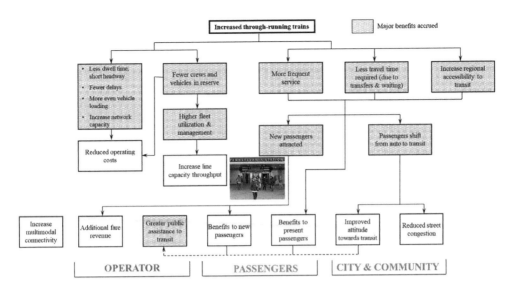

Figure 5. Network efficiency on through-running operations.

Equation 5 shows the minimum headway along the line, depends on the following elements: where standing or dwell time at station t_s.

Type of control and required safety regime, which affect reaction time t_r and safety K ($\geq$1); this factor is commonly used in rail transit operations to consider the conditions when normal braking rates cannot be achieved due to variations in braking rates, etc.

Length of TU ($n \cdot l$), which affects station clearing time of the train

Acceleration and braking rates, a and b, respectively, and speed of the arriving TUs, V (Bruun, 1992; Vuchic, 2005).

On a through-running station, trains are able to pass through the same station at a minimal headway ($h_{s\ min}$). Such minimal headway already considers the required operating safety interval given by the federal railroad authority. In contrast, on a dead-end terminal, trains are unable to pass through at a minimal headway ($h_{s\ min}$); in fact, each train has to operate at two times of the minimal headway ($2 \times h_{s\ min}$). Even worse, in an old-fashioned communication-based train control (CBTC) system, like the New York Trans-Hudson tunnel one, the operating safety (interval) for Amtrak is 15 minutes in each direction. Accumulated deadhead time on a single weekday, caused a profound economic loss to the New York metropolis.

The trade-off between fix tracks with lower investment but higher O&M cost vs. flexible tracks with higher investment but lower O&M cost has also been discussed. Based on time-distance measures on through-running, superior operating efficiency has been found with the through-running operations compared to dead-end operations. Figure 5 summarizes through-running' operation advantages across the operator, passenger, and city.

4 PENNSYLVANIA STATION: BREAKDOWN OF EXISTING CHALLENGES AND CORRESPONDING ENGINEERING MEASURES

Currently New York City' (NYC) regional rail systems and network are unable to accommodate the rising ridership demand. In Figure 6. Pennsylvania Station is a dead-end terminal located in the center of Midtown Manhattan. Regional rails include Amtrak, New Jersey Transit (NJT), Long Island Rail Road (LIRR), carry millions of daily commuters in-and-out of Manhattan. The network efficiency on operating flexibility, continuity of lines, and transfers at Penn station are low. Train movements at Penn station are inefficient due to the inflexible track alignments, resulting in low fleet utilization (New York Times, 2014; The Wall Street Journal, 2016). Penn station is also disconnected from Grand Central station; thus, disconnected with the Metro-North Railroad service, which links to the Northeastern part of the U.S. From the city's functionality and long-term competitiveness perspectives, transit is operating on a series of disjointed parts with limited regional connectivity. To turnaround, the city would require a transformative action to rethink its urban strategies: connect the disconnected parts and increase efficiency gain from an integrated network (Venturi, 2017). The RethinkNYC Plan 2050 explored the geospatial metadata: population distribution, O-D survey, transit connectivity, land use pattern, and future urban growth within the tristate regions, which includes New York, New Jersey, Connecticut. (The Architect Newspaper, 2016; Bloomberg Benchmarks, 2017). The construction phasing plan mapped a series of schematic station design on platform expansion, track reengineering, and network realignment at the underperforming Pennsylvania Station. This section breakdowns three major challenges and provides corresponding engineering measures to convert an independent dead-end terminal to an integrated through-running station at Pennsylvania Station.

4.1 Low Network Capacity vs. Flexible Track Alignment for Higher Operation

First, the current tracks are operating at maximum network capacity. The inflexible track alignments are impossible to accommodate extra services or any incidental changes. Following trains (FTs) have to wait in tunnels for 15 minutes as leading trains (LTs) exit the station. The station is operating as a terminal rather than as a through station. Trains must cross each other as they enter and leave the station as shown in Figure 7. (Left). The countermeasure of through-running avoids congestion by scheduling eastbound traffic on southern tracks and west-bound traffic on northern tracks shown in Figure 7. (Right). Each train would enter the station, prepare for passengers boarding and alighting, and continue without ever crossing incoming and outgoing traffic.

To enable the steady flow of through-running operation, a feasibility study of phasing plan incorporated with track reengineering, network realignment, and minimization of construction and demolition for normal operation are comprehensively examined (Venturi et al, 2017). The study identifies a total of 10 phases to convert the existing independent dead-end terminal to an integrated through-running station (Figure 8. Left). Each phase has its counter

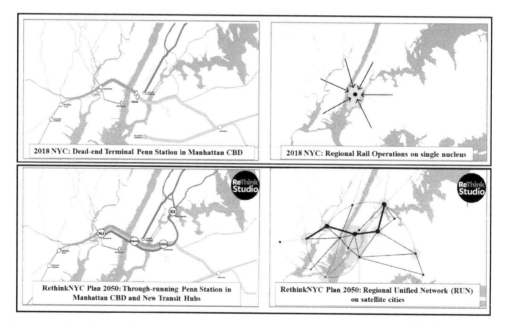

Figure 6.   New York Regional Rail Network: Discontinued (Up) vs. Unified Services (Down).

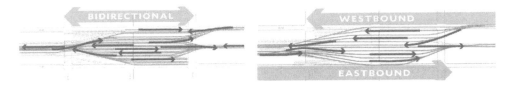

Figure 7.   NY Pennsylvania Station Dead-end Conflict (Left) vs. Through-running Flow (Right).

operating strategy to be followed (Figure 8. Right). The phase plan balances the construction and demolition timelines without disrupting normal commuting services.

4.2   Limited Passenger Circulation vs. Platform Expansion to Expedite Boarding and Alighting Process

Second, narrow platforms are considered as a safety issue. Overcrowding on platforms is mitigated through staged boarding, which forces outbound passengers to wait on the mezzanine level until passengers on the leading train are totally alighted. Limited vertical circulation (stairs, escalators) produces chaotic passenger flows and rushed transfers, especially to the NJT passengers (Figure 9. Left). Without even discussing the possibility of protective screen-door installation, overcrowding drastically reduces system reliability. In contrast, a through-running station would allow single-track configuration to expand the platform width and additional space for vertical circulation (Figure 9. Right). Such measure offers greater safety and increases passengers' boarding and alighting process at platform and mezzanine levels.

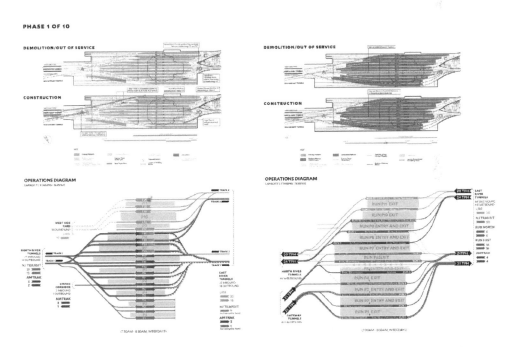

Figure 8.   Selective schematic design on track reengineering, network realignment (Left) and counter operation strategy (Right) within the Penn Station Construction Phasing Plan.

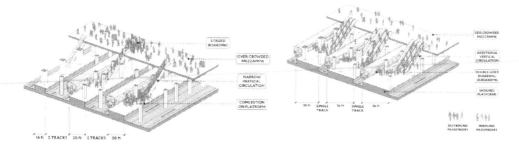

Figure 9. Penn Station existing platform condition (Left) vs. engineering improvement on vertical circulation (Right).

4.3 DISCONNECTED NETWORK SERVICES VS. UNIFIED NETWORK TO INCREASE REGIONAL CONNECTIVITY

Third, the New York regional rail services are disconnected. Different landmasses (Manhattan, New Jersey, the Bronx, Long Island) have different transit agencies. Passengers who would like to travel in between New Jersey and Long Island must experience Penn Station's narrow platforms and unreliable boarding and alighting process. To whom would like to transfer from Penn station to Grand Central must use MTA subway. To enable Penn station through-running's capability, the adjustment of terminal functionalities include relocation of railyards, reduction of long dwell times at platforms, and execution of the two abovementioned measures (*A & B*). Instead of operating a single nucleus terminal in the center of downtown Manhattan, the plan is to evenly distribute ridership by leveraging the gravity of satellite cities and incubating the growth in Port Morris, the Bronx and Secaucus (Previously shown in Figure 6), where belong to multimodal transit hubs.

The RethinkNYC Plan 2050 addressed the necessity to unify New York regional rail services. The construction phasing plan provided schematic adjustment on platform expansion, track reengineering, and network realignment at the current underperforming Penn Station to enable the through-running services to become a cohesive regional network. In Figure 10. tunnel capacity and service density have been increased, regional connectivity has been improved, and fleet utilization has been better performed.

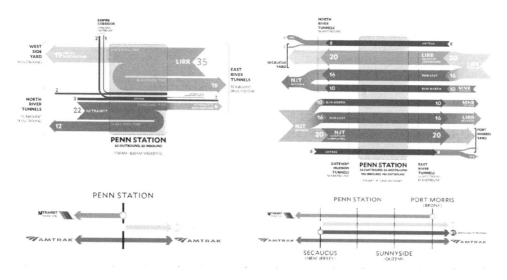

Figure 10. Comparative analyses of dead-end (Left) vs. through-running (Right) network capacity at the New York Pennsylvania Station.

5 CONCLUSION

The diversity among cities, metropolitan areas, and even countries in terms of their historic, geographic, social, and strategic positions dictate to the requirements for a variety of approaches to urban transportation problems. Policies and solutions cannot be directly transferred from one to another; however, many fundamental problems are similar, and the exchange of experience can be useful in resolving the sophisticated problems faced by cities.

Many global cities are either in the transition stage of repositioning cities' long-term competitiveness or in the development stage of a large-scale metropolitan planning. Common facts have been found in these two settings: the outstanding commitments on the modernization of efficient urban transit systems and the transformative mindset on the recapitalization of city's assets both developable and underutilized lands within the CBD and surroundings. Tunnel efficiency and capacity have a great impact on a city's functionality. This relation is obvious in the urban environment. The solution, proposed by the Rethink Studio, would modernize the inflexible Penn Station track geometry and transform the NYC regional rails from the hodgepodge of 19th century rail lines to a regional unified rail (RUN) system (Venturi, 2017).

Reengineering efforts from an independent dead-end terminal to an integrated through-running tunnel station resulted in a more efficient regional unified network. Extensive trade-offs to network efficiency have been given. Encouraging through-running operation and discouraging dead-end terminal engineering measures have been introduced. One of the planning challenges in the U.S. is that there are too many competing entities and each has its agenda. A simple method for streamlining these competing interests has not yet been found. Regional and national planning starts by reaching consensus on general guidelines about the type of city and society and then determining the composition of a multimodal system as well as specific plans for each mode.

U.S. cities, unlike the Chinese, Russia, and other centralized decision-making countries, encounter administrative contradictions and managerial barriers; however, a relentless efforts must be devoted to understanding of the roles and characteristics of different transportation modes, their impact on the long-term sustainable growth of metropolitan areas, and both tunnel and transit network efficiency on economic value creation. This needs to happen while avoiding mutually conflicting policies to achieve an intermodal balanced transportation system. A plan always comes with a purpose. A broad vision of the city-transportation relationships and the creation of regional unified networks are interdependent with counties' long-term competitiveness (Vuchic, 1999).

REFERENCES

Benz, G.P. 1987. Transit Platform Analysis Using the Time-Space Concept. *Transportation Research Record 1152*, TRB, National Research Council, Washington, D.C., pp. 1–10.

Bloomberg Benchmarks. 2017. *Why New York's Summer of Hell Matters to More Than Commuters.* Retrieved 22. June, 2017 from https://www.bloomberg.com/news/audio/2017-06-22/why-new-york-s-summer-of-hell-matters-to-more-than-commuters

Bruun, E. 1992. *The Calculation and Evaluation of the Time-Area Parameter for Any Transportation Mode.* Ph.D. dissertation, Department of Systems Engineering, University of Pennsylvania, Philadelphia

CBS News. 2017. *More Tunnels Would Greatly Ease Service Problems At Penn Station, Experts Say.* Retrieved 4. April, 2017 from http://newyork.cbslocal.com/2017/04/04/more-tunnels-would-greatly-ease-service-problems-at-penn-station-experts-say/

Musso, A & Vuchic, V.R. 1988. Characteristics of Metro Networks and Methodology for Their Evaluation. *TR Record 1162*, pp. 22–33, Washington, DC

New York Times. 2014. *Thinking Big. Then Thinking Bigger. An Idea to Restructure and Expand La Guardia Airport.* Retrieved 9. November, 2014 from https://www.nytimes.com/2014/11/09/nyregion/an-idea-to-restructure-and-expand-la-guardia-airport.html

New York Times. 2016. *Thinking Big and Bigger About New York.* Retrieved 25. March, 2016 from https://www.nytimes.com/2016/03/27/nyregion/thinking-big-and-bigger-about-new-york.html

The Architect Newspaper. 2016. *Jim Venturi and ReThinkNYC want to revolutionize how NYC handles train infrastructure*. Retrieved 26. May, 2016 from https://archpaper.com/2016/05/jim-venturi-rethinknyc/

The Wall Street Journal. 2016. *La Guardia's Runways Come Up Short*. Retrieved 13. November, 2016 from https://www.wsj.com/articles/la-guardias-runways-come-up-short-1479078957

Venturi, J.C. 2017. *RethinkNYC Plan 2050 Penn Station and A Post-Moses New York*, Cooper Union, New York, Retrieved May 9, 2017 from https://www.youtube.com/watch?v=U_-CoiFIb7A

Venturi, J.C. 2017. *ReThinkNYC Executive Summary*. New York, NY, 2017.

Venturi, J.C. et al. 2017. *The Regional Unified Network. Volume 1. First Step*. New York, NY

Venturi, J.C. et al. 2017. *ReThinkNYC Penn Station Through-running Construction and Demolition Schedule Operations*. New York, NY

Venturi, J.C. et al. 2017. *The Regional Unified Network. Volume 2. Plan 2050*. New York, NY

Vuchic, V.R. & Musso, A. 1991. Theory and practice of metro network design (in English, French and German). *Public Transport International*, pp. 298–325, Brussels.

Vuchic, V.R. 1999. *Transportation for Livable Cities*. New Brunswick, NJ: CUPR, Rutgers University.

Vuchic, V.R. 2005. *Urban Transit: Operations, Planning, and Economics*. Hoboken, NJ: Wiley & Sons.

Vuchic, V.R. 2014. *Planning, Design and Operation of Rail Transit Networks*. Metro Report International, pp. 48–53. London, UK.

Tunnels and Underground Cities: Engineering and Innovation meet Archaeology,
Architecture and Art, Volume 10: Strategic use of underground
space for resilient cities – Peila, Viggiani & Celestino (Eds)
© 2020 Taylor & Francis Group, London, ISBN 978-0-367-46878-1

Tunneling challenges in the business capital of India

S. Vishwakarma
Tata Projects Limited, Mumbai, India

V.K. Thakur
ITD Cementation Ltd., Mumbai, India

U.S. Virat
Mumbai Metro Rail Corporation Limited, Mumbai, India

K. Joshi
CEC- ITD Cem- TPL Joint Venture, Mumbai, India

ABSTRACT: Underground Tunneling has always been daunting when it comes to some of the oldest and busiest cities of the world. A lot of challenges await the construction team starting from the soil diversity to selection of most appropriate TBM's, maintaining safety of existing structures and residents of the area being the topmost priority. In the present scenario, demand of Tunnel Boring Machines has suddenly increased worldwide to construct the Metro tunnels in Metropolitan cities, but issues related to vibrations caused by use of TBMs to nearby structures have been a matter of concern. TBM sometime based upon the soil profile are the source of vibrations that causes distortion due to its impact and displacements of material under the foundation of structures.

1 INTRODUCTION

The problem of tunnel-induced stresses produced at the time of construction and related risk assessments of buildings damages has attracted attention of investigators over the last many years and so these all have to be taken into consideration and analyzed at all times.

In this paper, we will go through the on-site challenges faced during tunneling of one of the tunnel drives from Mumbai Metro Line 3(MML3) project considering few case studies.

2 PROJECT BACKGROUND OF MUMBAI METRO LINE 3

2.1 Geological Overview of Mumbai

Mumbai is located in the western coast of India, by the bank of the Arabian Sea. Back in the 16th century, Mumbai was known by the name "The seven islands of Bombay" because it is made from the cluster of 7 islands comprising of Isle of Bombay, Colaba, Old Woman's Island (Little Colaba), Mahim, Mazagaon, Parel and Worli. By 1845, the islands had been merged into one landmass by means of multiple land reclamation projects.

2.2 Regional Geology

Mumbai region is located in the great volcanic formation building up the Deccan plateau.

A wide variety of Basalts and associated rocks such as volcanic breccia, black tachylytic Basalts, red tachylytic Basalts etc. occur in the area covered by Deccan trap Basalts. Most Basalts are either compact i.e., with no gas cavities, or amygdaloidal with gas cavities filled with secondary minerals, and vesicular Basalts with empty gas cavities. Zeolites are the commonest secondary minerals filling gas cavities, though silica, calcite and chlorophacite also occur as infillings.

2.3 Need of the Project

Mumbai, the financial capital of India, has witnessed phenomenal growth in population and employment and the trend is expected to continue in the future. Four-fold growth of population since 1951 has been largely accommodated in the suburbs while the highest concentration of jobs has remained in the Island City. The physical characteristics of the City are such that the suburbs have been constrained to spread northwards only, and all transport facilities are concentrated within three narrow corridors. Today's major challenge is to provide connectivity and promote growth by providing additional infrastructure, which would improve the quality of life of the residents.

2.4 About Mumbai Metro Line 3

Mumbai Metro line 3 connecting Colaba- Bandra- SEEPZ is the first underground Metro line in Mumbai metro system having 26 underground stations and 1 at grade station. The total length of MML-3 is 33.5 kms and the total cost of the project is $ 3.19 billion with 57.2% soft loaned from Japan International Cooperation Agency (JICA) and rest 42.8% from Government of India/Government of Maharashtra/Others. It is further bifurcated into seven contract packages namely UGC01 to UGC07 and is implemented by The Mumbai Metro rail Corporation Limited (MMRCL). Once completed, Mumbai Metro Line 3 will cater to 1.39 million daily ridership by 2021 as estimated.

The case studies in this paper pertain to package UGC04 of MML3 corridor which comprises of 3 underground stations named Siddhivinayak, Dadar and Sitladevi and associated bored tunnels (cumulative tunneling length of 10992 meters), cross passages and NATM tunnels at Sitladevi station. The project is being executed by a Joint venture of Continental Engineering Corporation, ITD Cementation India Limited, and Tata Projects Limited (CIT JV).

2.5 Brief Description and Alignment of UGC 04 of MML3

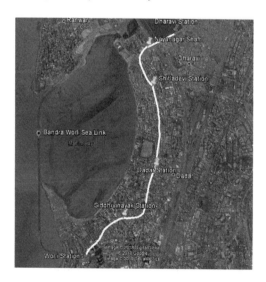

Figure 1.   Tunnel alignment for UGC 04.

Table 1. Project Brief.

Project	Mumbai Metro Line 3 – UGC 04
Client	Mumbai Metro Rail Corporation
GC	MAPLE Consortium (Joint Venture of AECOM, Padeco, LBG Inc. and Egis Rail)
Scope	Design & Construction of Underground Section including Three Underground Stations at Siddhivinayak, Dadar & Sitladevi & Associated Tunnels together
Location	Worli End to Dharavi Start
Tunnel	Inner Dia – 5.80 M
	Outer Dia – 6.35 M
	Segment Width – 1.4 M (Universal Type)
	Segment Thickness – 275 mm.
Stations	Siddhivinayak (250m L)– Bottom Up with secant pile
	Dadar (324m L) – Bottom Up with secant pile
	Sitladevi (240m L) – Bottom Up with secant pile

2.6 Geological profile along package UGC-04

Rock type Basalt (about 75%) and Volcanic Breccia and Tuff (about 25%) are encountered along the alignment with some amount of intertrappean shale (interbedded with Basalt or Volcanic Breccia). The ballast rock generally having high intact strength varies between 5 to 73 MPa. Rock is mostly within weathering grade II to III followed by grade I. Thin bands of tuff are present beneath the soil layer having a weathering grade of V and IV. The bored tunnel section consists mostly of Basalt in weathering grades I to III except in the tunnel portion after Nayanagar launching shaft where Breccia of grade I to III is encountered. Basalt rock does not have very high abrasivity, but locally intact rock strength could be high. Permeability values are generally low to moderate.

We were dealing with face conditions with rock mass varying from grade 1 to 4 throughout the alignment.

3 SELECTION OF TUNNEL BORING MACHINES

After the detailed soil analysis, selection of TBM was most important, as tunneling was going to be executed in the city that had undergone various land reclamation projects and any surprises may await the execution team, so several criterions were considered before selecting the appropriate TBM and its specifications.

Based upon the geology pertaining to UGC package 04, Earth Pressure Balancing Shield (Herrenknecht AG) was chosen as the best-suited TBM w.r.t. required progress and safety of Tunneling Operations & existing building structures (EBS).

Figure 2. Tunnel Boring Machine used in UGC-4.

3.1 Cutter Head design

The cutter head was designed with an opening ratio of 33% and for the rock strength of UCS upto 150 Mpa. This anticipated rock strength was based upon the available geological data from the tender document that was subsequently verified with respect to detailed geological investigation conducted in the stretch.

4 CHALLENGES WHILE TUNNELING

While we tried to cover all the known risks through the selection of TBM and design of permanent structures but we still faced the following challenges presented in the form of case studies:

1. Launching of TBM.
2. Tunneling below old buildings.

4.1 Launching of TBM

The major tunneling activities for this package were exclusively focused on the construction of the TBM launching shaft at a place called Nayanagar, in Dharavi area of Mumbai as almost 50% of the total tunneling of the package was planned to be executed from this shaft.

As launching of 2 TBMs and 4 drives were to be executed from this shaft, so its construction on time was one of the critical activities for the project. The following challenges were faced as briefly described below:

4.1.1 Land for TBM Launching Shaft Construction

The land selected for the shaft construction was amidst a densely populated slum area, Dharavi which is the 2nd largest slum in Asia and the 3rd largest in the world. 404 hutments were to be removed to clear the space for shaft construction.

For land acquisition and hutment removal all the families living in the areahad to be relocated; so a special resettlement and Rehabilitation policy for Project Affected Person (PAP's) was framed by the State Government.

4.1.2 Resettlement and Rehabilitation of Project Affected Person (PAPs)

i. State Govt. approved to make Mumbai Urban Transport Project Resettlement and Rehabilitation (MUTP R&R) policy applicable to the project PAPs GR dt.03 March 2014.
ii. Around 2800 families are affected by project.
iii. Social Impact Assessment (SIA) report was prepared by General consultants as per Japan International Cooperation Agency's (JICA) Socio Economic and Environmental guidelines in Jan 2016 after incorporate changes in impacts.
iv. Public consultation were conducted with all affected slum dweller families during year 2016.
v. Availability of tenements; Hon'ble Chief Minister had approved 4000 nos. of tenements for ML3 PAPs from Slum Rehabilitation Authority scheme at Andheri, Chakala, premier

Figure 3.   TBM Launching Shaft Area - Nayanagar.

compound Kurla (W), Bhandari metallurgy Kurla (E) and Mahul. SRA handed over 500 tenements at premier Kurla. SRA also handed over 34 tenements to MMRCL at Antop Hill Wadala.

vi. TBM launching shaft for package 4 of is constructed on a plot of 3355 sq. m by clearing encroachment 404 hutment/slum of Govt. Land.

vii. Baseline Socio Economic Survey (BSES) was conducted by General consultants as a part of SIA to decide the eligibility for Resettlement and Rehabilitation of Project Affected Person (PAPs) on the said plot at Nayanagar.

viii. After detailed scrutiny of BSES forms and SIA report list of eligible PAPs was prepared by R&R cell of MMRC and accordingly public consolations was conducted at MMRC office with Eligible PAPs.

ix. Alternative options were discussed with PAPs regarding location of resettlement, residential commercial etc. and PAPs agreed for Premier Kurla Resettlement sites.

x. Tenements were allotted to eligible PAPs through Computerized lottery system.

xi. After collection of all requisite documents, allotment letters were issued to PAPs for tenement and 1-week time was given for vacating the hutments at Nayanagar in October 2016.

xii. Transportation arrangement were provided by MMRC/Contractor for shifting of home luggage of PAPs to Kurla premier site.

4.1.3 *Space Constraint for TBM Operations*

Figure 4.   Shaft Area after TBM Arrival.

As 2262.42 sqm (Excluding the Shaft Area) space was available for setting up of two TBM back up services like Grouting Plants, Cooling Towers, Mucking Gantries, Muck Pits and TBM Power supply which was a challenging task w.r.t. movement of heavy machineries and cranes etc along with maintaining the planned tunneling progress.

This challenge was tackled by carefully planning the gantry crane setup for parallel operations of two TBMs and optimized utilization of the available land.

4.2 *Conclusion*

The first ring for TBM1 (named as Krishna 1) was erected on 06 November 2017 beginning Initial drive of the first TBM of Mumbai Metro Line 3 from Nayanagar Shaft towards Dadar Station, in the presence of Honorable Chief Minister of Maharashtra Mr. Devendra Fadnavis. It is also notable that UGC04 has the longest stretch for tunneling among all the seven packages of MML3 and so timely start of tunneling drive was necessary.

4.3 *Case Study 2: Tunneling below Old Buildings/Dilapidated Buildings*

This case study takes few examples to define the challenges that were faced while tunneling below some of the dilapidated buildings and the major steps taken for safely executing the project.

Many buildings in Mumbai are currently unfit as a habitat. Some of such buildings in the influence zone* of our TBMs and Station area, have to be analyzed for the impact of the construction activities. Therefore, a pre-construction condition survey was carried out and based upon which a protection plan consisting of monitoring and structural support plan was framed for these buildings.

*Influence Zone: *The 50 meters distant area parallel to the TBM centerline on both sides is defined as the Influence Zone of the TBM.*

4.3.1 Pre-Construction Condition Survey

After conducting the Pre-Construction Condition Survey of all the buildings of the entire stretch lying in the influence zone, the reports were submitted to the DDC (Detailed Design Consultants for Contractor) for analysis and categorization of buildings was done according to the degree of estimated damage to the buildings. After studying the survey reports of all the buildings, it was concluded that most of the buildings did not require any structural repairs and were competent to bear the effects of tunneling activities but for the dilapidated buildings, special provisions were made to take care of the buildings and their residents.

4.3.2 Building Protection Plans

Structural audits were also conducted by an MCGM (Municipal Corporation of Greater Mumbai) authorized agency and reports were submitted to DDC for incorporating the necessary provisions to protect the buildings in the building protection plan. Some of the special provisions included:

- Structural repairs of various structural members of the buildings.
- Real-Time Monitoring.
- Increasing the frequency of monitoring of buildings in the bubble zone of the TBM.

*Bubble Zone: *The zone lying 100 metres behind the TBM and 50 metres ahead of the cutter head.*

4.3.3 Tunneling below Dilapidated buildings (St. Xavier Institute of Engineering)

The problem with tunneling in Mumbai under dilapidated buildings was that for many of the buildings the as built drawings of structures were not available and even if it was available, the actual geometry of the buildings had been altered by various repair works undergone over due course of time. As the tunneling activities started, our upline TBM had to pass below 2 such buildings just 110 metres away from the Nayanagar Shaft; St. Xavier's institute of Engineering (G+3; slightly damaged) and St. Xavier's Technical institute of Engineering (G+5; Severely damaged)

4.3.3.1 ST. XAVIER'S INSTITUTE OF ENGINEERING

The first building we encountered lying above the Tunnel alignment was St. Xavier Institute of Engineering (G+3).

- The building was assessed to be in category 2 (Slightly damaged) according to the BCS Report.
- In addition, the impact of Tunneling over the building was also found to be in category 2 (Slight Impact) by the Construction Impact Analysis Report.
- The architectural and structural details of the buildings were unavailable.
- At first, it was inferred that it has isolated RCC footings of depth 2.5 metres.
- Accordingly, the DDC prepared a monitoring plan considering the foundation to be isolated RCC footing.
- Further deeper research with the owners and local body it was confirmed that the building is resting on pile foundation.

After this revelation, a joint meeting among the stakeholders was conducted to carry on necessary research and analysis.

Initially we could not find any drawings for the building foundations but after digging in deep a blue print was obtained which had details about the piles of the building and it proved out to be very critical factor in analyzing the tunneling impacts on the building.

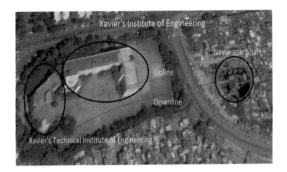

Figure 5.   Satellite view of Nayanagar Shaft and St. Xavier Institute of Engineering.

(1) Preliminary analysis:
  • Preliminary analysis included Ground settlement analysis, Analysis of loads and Bending Moment and Building Condition Survey (BCS).
  • Construction Impact Analysis and Building Damage Assessment were to be done for any concrete outcomes regarding the piles of the building.
  • Numerical Modelling of the building Structure was conducted by DDC to analyze the Ground Settlement and loads on the structure.
  • Monitoring of the structures using various instruments at regular intervals started.
(2) Further Investigation of building:
  • After analyzing the pile layout, it was observed that the four columns along the edge of the building were coming above the tunnel alignment.
  • It was mentioned in the pile drawings that the pile shall be embedded into the hard rock/stratum to a depth equal to the diameter of the pile. Any deviation in length of embedment shall be with the specific instructions in writing from the consultant.
  • The pile foundation can become sensitive to the effects of tunneling. Piles risk a reduction of their end bearing capacity and skin friction resistance due to the displacements and ground stress redistributions that occur as a result of tunneling.
  • Thorough study of the substructure and superstructure was conducted.
  • The cross section across the tunnel alignment at the location of the building is modelled in RS2 (A Finite element software).
(3) The following reports were generated regarding the assessment of the structure.
  a) Ground Settlement Analysis Report:
    Construction impact assessment is done considering the ground movements at 13m below ground level where the pile termination is expected. Whereas, the building will

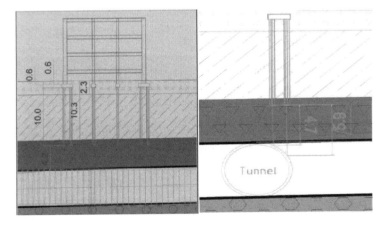

Figure 6.   Piles of St. Xavier Institute of Engineering with respect to the Tunnel Alignment..

not undergo settlement to this extend since the skin friction along the pile will be mobilized and the vertical settlement is restricted.

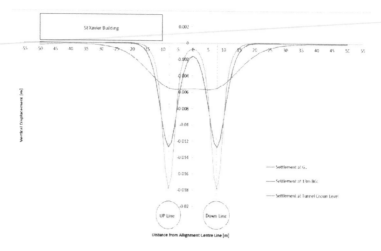

Figure 7.   Ground settlement description.

b) Analysis of Bending Moment:

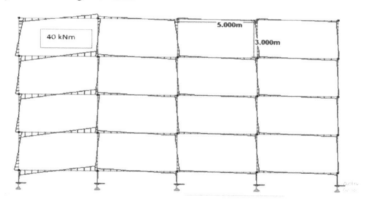

Figure 8.   Bending Moment Description.

The STAAD analysis of one bay of the building with differential settlement obtained from Finite Element Method at pile toe is applied as load which shows that the building is capable of sustaining the additional forces coming on it. Also, the analysis shows that the bending moment produced in the columns is negligible and can be dealt by the skin friction of the pile and the rock surface.

c) Aftermath of the research done by DDC and CIT JV:

The following conclusions are drawn from the calculations and submitted designs: -

1. Based on the data from additional probe holes conducted near the institute building, rock is encountered at a depth of 11m to 12m from the ground. Rock is considered from 12m depth for the pile capacity calculations.

2. From the pile capacity calculations, it was noticed that the pile load was well within the pile capacity even after ignoring end bearing capacity of the pile.

3. It was noticed from Construction Impact Analysis that the expected settlement of the ground at the level of pile toe due to the TBM operation does not impact the structural integrity of the structure since it falls in slight category.

4. None of the piles were coming directly over the central axis of tunnel which is the highest point of boring. In fact, the pile away from the tunnel axis and towards the building side would have an even larger cushion of rock available between the pile toe and tunnel.

5. It should be noticed that these calculations are carried out for the two center columns. The piles for the edge columns will have an even lesser load.

6. It should also be noted that the calculations are done for a pile length of 13m where it terminates above the tunnel crown. Even if the pile extends beneath the tunnel crown, the skin friction alone will be sufficient to take up the load on to the pile even after applying the reduction factor. However, extreme caution should be taken during the TBM drive due to the close proximity of piles to the excavation and the uncertainty involved in the length of piles.

d) Recommendations:

The following recommendations were made in a joint meeting between the stakeholders after studying the results of the conducted tests to safely carry out the task: -

1. First 2 bays of the columns (approx. 15m from the edge of the building) near the building edge close to the tunnel alignment were evacuated. This was a preventive measure.

2. It was recommended to advance the TBM with slower rate and with closed mode to minimize the settlements.

3. Control of TBM face pressure – in order to minimize the settlements near the pile foundations of the building ahead of the cutter head, it was recommended to increase the face pressure a slightly higher (20-40 kPa) than the recommended face pressure. The recommended face pressures during TBM drive at the location of St. Xavier Institute of Engineering are as given below:

   At tunnel crown – 130 kPa

   At tunnel center line – 163 kPa

   At tunnel invert – 196 kPa

4. As a contingency measure, it was recommended to stack the props at site and readily available for propping if required

5. It was recommended to install 2 numbers of vibration sensors near the column locations of the building at ground level for the vibration monitoring due to TBM drive.

6. It was recommended to install some additional ground settlement markers along the alignment of tunnel centreline and also between the location of building and current TBM location for monitoring of surface settlements

7. The instantaneous filling of the "annulus", which is created behind the segment lining at the end of the shield tail, was an operation of paramount importance. Its main goal was to minimize surface settlements due to the unavoidable over-excavation generated by the passage of the TBM. Therefore, it was recommended to carry out the tail grouting soon.

8. It was recommended to install 2 numbers of additional 3D optical targets on the building face towards tunnel alignment.

9. The frequency of the monitoring should be increased to every 2 hours. Any visible sign of distress in the building should be monitored and reported. The field instrumentation and monitoring reports shall be prepared and circulated to DDC/GC Design team on daily basis.

10. The vibration at the Ground Level due to the TBM operation shall be limited to not more than 4 mm/s.

11. The building should be monitored continuously using settlement markers, crack meters, tilt meters and 3D targets as per the submitted drawing and the additional instruments recommended as mentioned above (points 5 and 8).

12. Extreme caution should be taken during the TBM drive in the vicinity of the building due to the close proximity of piles to the excavation and the uncertainty involved in the length of piles.

13. Forward probing from TBM should be carried out to confirm the absence of piles.
e) Monitoring done by Instruments at site:
The following Instruments were installed at St. Xavier buildings:
a) Crack Meters
b) Ground Settlement Markers
c) Building Settlement Markers
d) Tilt Plates
e) Vibration Meters
f) Stand pipe Piezometer
g) Bi- Reflex Targets
h) Inclinometers
i) Multi Point Borehole Extensometers (MPBX)
(4) Propping at vulnerable locations:

Figure 9.   Supporting Arrangements installed inside classrooms of the St. Xavier's Technical Institute.

4.4   *Conclusion*

The St. Xavier Institute building was the first uncertain building whose construction data was not available. Since the building's damage category was slight it was not posing a threat but continuing without ascertaining the pile location was never an option.

Similar analysis was conducted for the 2nd building i.e. St. Xavier Technical Institute of Engineering which was much more critical than the first building and it required extra prop supports considering the critical nature of the building.

Therefore, in the absence of proper foundation details, tunneling was carried out by proper analysis of the ground conditions, impact analysis, strict instrumentation monitoring, perfect operations and hence tunneling was completed without any damage to the sub structure or superstructure of the buildings.

Tunnels and Underground Cities: Engineering and Innovation meet Archaeology,
Architecture and Art, Volume 10: Strategic use of underground
space for resilient cities – Peila, Viggiani & Celestino (Eds)
© 2020 Taylor & Francis Group, London, ISBN 978-0-367-46878-1

Compound use of underground space for integrated campus-urban development: A case study from Tsinghua University, China

W.J. Zhu
Department of Civil Engineering, Tsinghua University, Beijing, China

Q.C. Bao & S.J. Gai
Construction and Campus Planning Office, Tsinghua University, Beijing, China

ABSTRACT: Shortage of existing land resources and limitation of boundary expansion become challenges on campus in China. Rational use of underground space on campus and surroundings becomes an important trend for sustainability in traditional and new built campuses. As a famous traditional campus in China, Tsinghua University set up a research project on the use of underground space on its main campus. Developing and integrating the key massif and the underground space can effectively increase supply of stock space resources, avoid additional contradictions caused by the aboveground construction alone, and improve campus functions. Responding the planning of city transportation of which an underground road and Metro line 15 of Beijing city will pass the main campus, a concept of underground space use inside the campus connecting the urban underground traffic line and campus garages is developed, which will relieve the internal and peripheral maneuvers of the main campus of Tsinghua University.

1 INTRODUCTION

Tsinghua University (THU), as one of the most famous universities in Mainland China, is facing the challenges of campus land shortage, environmental improvement, traffic management and the coordination between campus-city spaces. The Geo Engineering Institute of Tsinghua University I commissioned by THU's Construction and Campus Planning Office to undertake the research on underground space development and planning under the main campus of THU. The objectives of this project were to:

- provide a conceptual plan and feasible proposals for underground space development on the main campus as an integral part of the THU' campus master planning.
- provide several examples of key projects in the near future of underground space development on campus, including implementation solutions.
- explore the feasible proposal of underground space utilization to meet the needs of the campus itself, as well as to develop the integrated collaborative solution with the main campus surrounding areas according to the Beijing Urban Master Planning and at meanwhile to respond the policy of space volume reduction requirements for urban land use of Beijing City.

This paper will describe the general idea and conceptual scheme of the integrated development of campus underground space (CUS) based on this special research project. The CUS Planning research conforms to the feasibility of integration development of campus-city space that the Beijing Urban Master Planning requires through three-dimensional composite utilization, which can improve the quality of traffic and public space on campus, and the overall traffic of the northwest part of the Beijing main city.

2 BACKGROUND AND CAMPUS OVERVIEW

2.1 *Location, geology and topography*

Tsinghua University is located in Haidian District, the northwest of Beijing, between the Fourth Ring Road and the Fifth Ring Road, of which the main campus covers an area of 306 hectares.

The region of THU campus is located in the alluvial and aggraded flood plain of Beijing area, which belongs to Quaternary strata with thick soil stratus where the bedrock buried depth is below 50–100 meters. The soil layer is mainly composed of sandy pebbles, local soft clay and artificial fill, and the bearing capacity is generally good, which is feasible to construction of underground space. The campus and its surrounding terrain are flat, generally high in the South and low in the North, of which the altitude varies at about 40 meters up to down. The Wanquanhe River, a natural river, which runs through in Haidian District, passes across the middle of THU main campus. After dredging and artificial hardening, the depth of the concrete floor of the river is about 8 meters. In the western part of the campus, there are several small lakes such at as the Jinchunyuan Garden and Tsinghuayuan Garden, which are famous for the Lotus Ponds. Historically, surface water was abundant on the campus areas, and there was a large area of rice cultivation. After a long-term development, most of the ground has become a man-made area, and the groundwater has been significantly lowered. However, compared with the urban center areas of Beijing city, the groundwater level in THU campus is still higher, generally 5–10 meters below ground, which suggests specific attentions for underground space construction.

Besides, the gardens and buildings left over from the Qing Dynasty, as well as the buildings in the early days of Tsinghua University, have been list as the objects of historical heritage protection, which demands stricter restrictions on the space expansion and the function perfection of the campus. Moreover, the large number of modern buildings cover most of the feasible lands for construction and renovation of the THU main campus. Fortunately, many open lands such as green fields and sports grounds provide potentials and specific advantages and opportunities for campus construction and renovation through development of underground space.

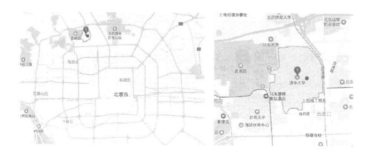

Figure 1.   Location and surroundings of THU main campus. (map.baidu.com.)

2.2 *The existing campus configuration and master planning*

The present main campus of THU can be recognized as two main parts, the western part newly established before 1949 and the eastern part newly established after 1949. The boundary of the whole main campus is surrounded by urban roads, adjacent to the Qing Dynasty imperial garden Yuanmingyuan Park, Chinese Academy of Sciences, Peking University, China Forestry University and Wudaokou Business Circle as the regional commercial center.

Tsinghua University began as Tsinghua School in 1911 established in the Qing Dynasty imperial garden named Tsinghuayuan Garden (initial named Xichunyuan Garden) at the western part of the present main campus. An American architect Emil Sigmund Fisher in China probably finished the campus's planning and designed the first educational buildings at Tsinghua in 1909, used a variety of foreign styles. (Zhu & Miao 2011)

In 1914, American architects H. K. Murphy and R. H. Dana completed the second master planning of Tsinghua School, which put the auditorium as the school center and expanded the campus to the west – the Jinchunyuan Garden, for University Department. The famous four important early buildings, namely the Auditorium, the Science Hall, the Gymnasium and the Library, were successively built and became the campus symbols of THU. Until 1928, the campus covered more than 80 hectares of land areas. During 1925–1928, when Tsinghua School was restructured into National Tsinghua University, the demand for campus scale and facilities expanded. Yang Tingbao, a Chinese architect as THU alumna, finished the third master planning of Tsinghua in 1930. The planning strengthened the spatial layout of Jinchunyuan Garden and the original campus, and expanded the campus to the north and south for residential and dormitory areas. According this planning, a large number of major campus buildings, such as the Biology Hall, Chemistry Hall, Meteorological Station, expansion part of Gymnasium and Library, Mechanical Engineering Hall, Electrical Engineering Hall, Hydraulic Engineering Hall, Civil Engineering Hall, Aviation Hall, and the West Gate of THU, had been built till 1937. The areas of THU campus covered 144.4 hectares and formed the early configuration of the western region of the main campus. (Luo 1984)

After 1952, according to the national adjustment of institutions for higher educations, THU became a comprehensive university with engineering as its main body. During 1954–1960, the new campus master plan was completed, which moved the old Beijing-Zhangjiakou Railway eastward 800 meters, and emphasized the Eastern Teaching Areas; meanwhile, the student and staff living areas were expanded to north and southwest. Total area of the planned campus expanded rapidly, reaching about 209 hectares, and forming the Eastern Part of the campus structure and the overall final outline of THU main campus. This plan emphasized rational zoning, compact land use, and increased the proportion of teaching rooms and sports venues, and made travelable roads and greens into a network. By 1966, the Staircase Classroom, Second Teaching Building and New Hydraulic Hall had been built in the old west district, and the Precision Instrument Building, Engineering Physics Hall, and the Main Buildings in the new east district. Meanwhile, the apartments and service facilities for staffs and dormitory buildings No. 1–12 for students were built in the southwest and the north district. (Luo 1984, Zhu & Miao 2011)

From 1979 to 2003, the recent campus master plans with better inheritance and continuity was completed by multiple revisions and the overall layout of two teaching central areas were defined in the west and east districts of the main campus. Meanwhile the concentrate construction of public areas of teaching buildings, the large-scale residences for staffs in the southwest and the large-scale dormitory areas for students in the northeastern, as well as the eastern playground and sports facilities were built. However, the campus function zonings are too far away to result in walking inconvenience and other issues, while the higher central-ized and largest teaching buildings group in THU campus leads to bicycle congestion and parking difficulties. From 2011, with the target of building world's top university, THU establishes a sustainable campus goal. Consulting report by American architect Venturi (2005) suggests that THU campus should enhance spatial comprehensiveness and diversity, combine classroom with diversified extra-curricular social education, supply easy linking and blending between disciplines, and make planning based on human needs and activities, as well as promote the continuity and practicability of green space and sharing of campus and urban space.

2.3 Campus transportation and urban planning in the region

The size of early campus of Tsinghua University was moderate and the campus buildings were relatively centralized. It was convenient to connect the teaching and living spaces by walking, and not related to the city outside the campus frequently. After 1954, with the rapid growth of university, the master planning of THU began to expand eastward and southward. Professor Wu Liangyong suggested the main gate of campus be located at the intersection of Chengfu Road and southward to Xizhimen district of Beijing City. In

1960, the plan of East-West and North-South main roads was formulated, and the road network on eastern campus was added to the total spatial structure of main campus. The route of Beijing urban planning road crossing THU campus along the East-West main road was reserved, and the long-term idea of the main road leading to Xizhimen in front of the Main Building of THU was put forward. This left a curse for the subsequent conflict between urban traffic and campus space (Zhu & Miao 2011). The Campus Master Plan of 1979, in order to solve the contradiction of urban bus stops over the distance, the possibility of urban planning roads across the east-west trunk road was remained. After 1988, with the campus space expansion of northward, a new main road crossing the east-west direction of the campus passing the Chemistry Hall was determined. The original route of Beijing-Zhangzhou Railway was determined as the North-South main road of the campus, and other two North-South main roads on both sides of the Main building were added. The general road network of THU's main campus was conformed but lack of separate consideration for walking and bicycle (Zhu & Miao 2011).

With the THU campus construction and developing, population and land use of Beijing City began to expand after 1952. After 1988, the number of motor vehicles outside and inside THU main campus accelerated, and the urban roads around the campus border were continuously densified and improved. In 1994, the Beijing municipal proposed the urban road crossing the east-west direction of the main campus from underground, for protection of campus environment, and the North-South Road in front of the Main Building should be reserved as municipal land for burring public pipelines, and the above ground space to be used and managed by THU (Zhu & Miao 2011). In 2008, the Beijing municipal proposed a plan for the co-operation of subway line 15 and the original planned underground road across the campus, but the co-construction has not yet been implemented. After 2014, the Beijing-Zhangjiakou high-speed railway and No. G7 National Expressway were planned and entering construction period. Up to now, Yuanmingyuan East Road, Heqing Road and Chengfu Road have formed an encirclement around the THU campus. Meanwhile, the urban roads of Tsinghua East Road, Tsinghua West Road, Zhongguancun East Road are vertically with and interrupted by the boundary of THU main campus. The congestion of motor vehicle around THU campus gates and the surrounding urban roads are on the top severe positions in Beijing. Additionally, the contradiction between vehicle and pedestrian and insufficient parking space inside and outside THU campus also influence the campus environment seriously. From 2011 to 2018, THU built two large residential areas for faculty and staff in Dongshengxiang District, east of the campus outside, which make a new demand of traffic links with the main campus for externalized sustainable campus.

Traffic inside campus and linking between internal and external campus and the integration of urban space combined with campus, as well as the potential concomitant impact on campus environment and the strategy are the important issues for both THU main campus and Beijing city. The effect of conventional management methods is limited. Integrated innovation of planning and management for the THU main campus combined urban space are necessary.

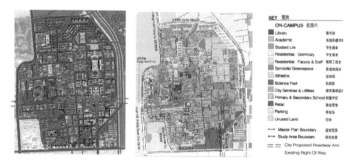

Figure 2.   Configuration and zoning of THU main campus.

3 RATIONALE FOR UNDERGROUND SPACE USE ON CAMPUS

Underground space planning of THU's main campus began with the compilation of the new version of the future campus master plan (2021–2035). Universities are places where the talents of national and social concentrates and carry on the knowledge inheritance and promotes thinking and innovation. The campus basic configuration are to provide necessary space and suitable environment to support and guarantee the effective conduct of academic activities. In a campus with large space scale and dense teachers and students, while providing enough space for study, experiment, practice, movement and growth, it is important to meet the requirements of a peaceful, convenient, free, communicative and communicative space atmosphere to support and nourish the learning and research activities and promote the growth of talents.

When the aboveground space are strictly limited, underground space becomes as the important part of land resource for alleviating contradictions and problems of land use, traffic and environment on campus. In the new version of master plan for THU main campus, combining with special projects on optimization of resources stock, traffic, landscape and construction of characteristic public space, the following points suggest the demand for underground space development on the main campus. Figure 3 shows the diagram of potential demand and locations.

1) There is a gap between the supply capacity of space resources on the existing campus and demand of land used to add necessary buildings and green landscape for future campus target. Apart from properly easing of some spaces that does not need to be inside the main campus, it is also necessary to expand some spaces to underground which brings less impact on environment without competing to the green space and let alone the capacity of increasing green space.

2) Parking demand in THU campus is large with many important areas crowded. Meanwhile, there are many places of conflicting between vehicles and pedestrian. In addition to traffic demand management to reduce the intensity of motor vehicle use, the setting of parking and driving to underground can reduce vehicular impact on the aboveground environment which help to create a car-free campus, and improve aboveground environment and walking continuity.

3) Campus zoning is obvious which makes single function, spaces far away each other, lack of public and communication space for students. In built up areas, development of underground public space by utilizing existing green space and square, can be used as places of study, communication, association, sports, investigation and experiment and other activities, and can supplement necessary places and suitable conditions of training and promotion for students while maintaining the aboveground environment and space by improving and connecting with each other. We can also solve problems of single function, because of insufficient comprehensiveness and lack of additional sites, and alleviate problems of long distance, inconvenient connection with functional zonings and new land use.

Figure 3. Sketch of underground space demand for solving contradiction on THU main campus.

4) Urban roadway with heavy traffic flow passing through the underground of THU campus can avoid environment impact on the campus ground space and avoid the large-scale campus obstruction to the urban traffic.

4 POTENTIAL & RESERVATION OF CAMPUS UNDERGROUND SPACE RESOURCE

4.1 *Suitable recourse for underground space use in the main campus of THU*

THU campus is located in a flat urban area, shallow geology is soil structure, and groundwater level is not deep. Therefore, the construction methods suitable for the underground space development of in the near future are the open excavation or mined excavation at shallow stratum, using cut and cover and shield or artificial NATM methods. For the architectural form, one is the attached basement of normal above ground buildings; the other is at the open space for the use of independent underground buildings. Depth of excavation can reach 20–30 m below ground, with about 3–5 floors of underground space which could be used as garages, tunnels, various laboratories, public spaces and storehouses.

4.2 *Suitable land use for underground space use in the main campus of THU.*

– Underground space development projects combined with surface renovation. According to the special investigation, there are 14 campus renovation projects suitable for underground space use, covering an area of 165,000 m².
– In combination with the protection and preservation of green space and squares, parking lot and playgrounds, independent underground space project can be added. After investigation and measurement, there are 19 open space plots suitable for independent development of underground space, covering an area of 130,000 m², and 17 playgrounds, square and parking lots, covering an area of about 256,000 m².
– Combining with the function improvement and upgrading of functional zoning areas, it is suitable for the expansion of underground space in the built-up area. There are two potential areas with value of research for future demand, which are the core area of Public Teaching Buildings and the area of front square and surrounding green axis of the Main Buildings.
– According to the continuity requirements of urban traffic or campus traffic, the potential underground space rout is remained on Tsinghua Road and other suitable roads for motor vehicles and underground land connection. According to the research, Tsinghua road and its green belt are the main resources for developing underground space road. In addition, School Road, e.g. Xinmin Road, Mingde Road, Zijing Road and Zhishan Road should be reserved for long-term underground transportation and municipal corridor construction.

4.3 *Controlling procedures for potential underground space developing plots in THU campus*

– It should be the first step to wake technical feasibility assessment for the potential underground space plots, which and when it meets any demand of future campus.
– It should be the second step to take ethics assessment for the technical feasible project of independent or high-density underground space development.

5 CONCEPTUAL PLANNING FOR UNDERGROUND SPACE USE

5.1 *Types of facilities to be located underground*

Appropriate underground space below the THU campus is envisaged to be in the forms of basements, independent underground architectures and tunnels with functions of transportation, municipal, civil air defense, various public spaces, scientific experiments and warehousing.

Underground space in this plan should be focusing on the function and space improvement coupled with the local land use conditions: 1) Intensive construction for increasing the volume ratio and green space area; 2) Improvement of spatial quality for existing areas which lack of specific service functions, such as increasing connectivity and public space without reducing green area. 3) Optimization of spatial stratified utilization for urban traffic and Open Campus – traffic access and connection channel. The typical applications envisaged for each type of space could include:

– Basements combined with aboveground renovation and new built projects: garage, storage, equipment rooms, laboratory, library, office, conference room, etc.
– Independent underground architecture projects for supplementing demand for existing and proposed extension projects. The Bauhinia Playground, East and West Playground, Front of Main Building, Art Museum Square, Teaching Building Core areas, Tsinghua Road, Affiliated Middle School, are the main preserving areas, for functions of student center, parking, indoor sports playground, public and communication space, display, supermarket, entertainment, laboratories, warehouses and civil air defense works, etc.
– Tunnels for vehicle and pedestrian traffic and linking of integrated underground areas: the core area of teaching buildings, the front area of Main Buildings.

Figure 4.   Sketch of typical independent renovations in underground and tunnels on THU main campus.

5.2   *Conceptual layout of compound campus underground space with urban transportation*

According to the urban planning of Beijing city, Tsinghua East Road will be extended westward through the underground space along Tsinghua Road inside THU campus and connected to Tsinghua West Road on the aboveground outside the West Gate of THU. Meanwhile, MTR Line 15 will be extended westward from the existing Tsinghua East Road Station, and will be transferred to MTR Line 4 at Yuanmingyuan Garden and outside West Gate of Tsinghua University. Tsinghua University can choose to set up a subway station on campus. As a result, the underground motor lane road will be constructed in the same way as the subway section. Due to the construction of Underpass Road and subway, although long-term environmental impact is less than that of aboveground road, it will still cause considerable interference during construction, as well as vibration and maintenance and management problems during the operation of transport facilities.

However, THU authority lacks adequate expectations and estimates of its own benefits and costs. Therefore, THU's enthusiasm for the construction of underground roads and subways is not high, in a status of waiting and hesitation. At the same time, the vehicle travel and parking demand along Tsinghua Road on the campus is very strong, but lack of parking space and effective countermeasures. Therefore, integrated relationship between urban roads and subway crossing the campus and the campus traffic and parking, and clarification of the relationship between THU and the surrounding urban space in terms of traffic, land use and environmental improvement is one of the key tasks in this research. The authors put forward the following ideas of integrated planning of underground space along the compound development line.

1) Supporting urban road plan from east to west across the underground space of THU campus. Along the Tsinghua Road with the underground road, combined with the parking demand of the campus, we plan four underground garages to be built. The urban underground road is required to establish a direct link with the four underground garages of campus, and the garages are still set up entrances and exits in the campus. In this way, urban transit vehicles pass through the underground road directly through the THU main campus, avoiding previous routes around the campus; and campus vehicles and visiting vehicles can enter the campus directly through the underground garage and parking, enter the workplace or carry out visits. Special and authorized vehicles can run through campus entrances and exit into campus aboveground space. Vehicles leaving the campus can leave the underground garage directly or the above ground vehicles leave the campus through the underground garage then enter the main traffic arteries of city.

2) Supporting metro line 15 across the campus to connect other lines to form a network. The line with underground motor lane is located below the driveway. The site and conditions of the subway station at the four corners of the School Road should be retained where subway station can be built temporarily or in the future. If a station is to be built, it is necessary to consider the management mode and feasibility of direct admission from the station, and the potential negative impact of different opening modes on the reputation of THU open campus. This plan proposes a possibility to construct an underground pedestrian corridor for the public to connect the Tsinghua Science and Technology Park 600 meters away. It is the main destination for people outside Tsinghua University to travel freely after arriving at the THU subway station. Reservation personnel can enter the campus interior space from the subway station.

3) In the West Gate area of THU, combined with the construction of transfer space between Metro Line 15 and Line 4, we suggest the underground public pedestrian corridor connected with the inner underground space of the West Gate so that the university people can swipe their cards directly into the West Gate. This can reduce the density and crowding of aboveground and the influence of crosswise pedestrian flow on vehicle traffic.

4) This research boldly proposes new planning proposals for adding urban underground roads to the campus. That is to induce the Zhongguancun East Road into the campus through underground space, to enter the underground space of the green axis-municipal pipeline land in front of the Main Buildings which was originally planned to be reserved as municipal road land. This underground part of Zhongguancun East Road converges in front of the Main Building to form a T-shaped underground road with extended Tsinghua East Road, where vehicles can be interchanged and transformed. In this way, the obstruction effect of large-scale campus space of THU on urban traffic can be greatly alleviated, and the road traffic efficiency of Northwest Cultural and Educational District of Beijing city, can be improved, which is expected to greatly alleviate the traffic congestion in this region. At the same time, it is suggested that the subsurface space above the underground Zhongguancun East Road as the land for repairing the campus public space in the front areas of Main Building. An underground complex with 3 floors and more than 20,000 square meters should be developed for

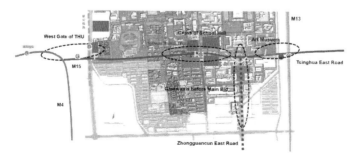

Figure 5.   Sketch plan of urban road and subway line 15 passing underground space of THU main campus.

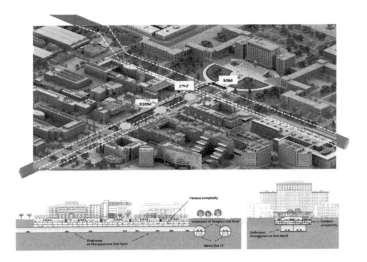

Figure 6. Proposed plan of urban underground road and subway line passing at Main Building areas.

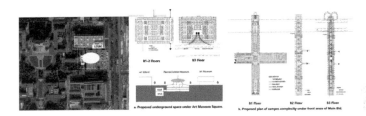

Figure 7. Proposed plan of two underground campus complexities connecting urban underground road along Tsinghua Road.

the campus parking, communication and display and service space in the southeastern campus region. The aboveground maintains green landscape, increases the local sunken square and entrance which can improve the original good landscape environment because of the visual effect is slightly flat, the green space public space use efficiency is not high regret so as to improve the transportation, walking and supporting in the front area of Main Building, and enhance the campus vitality.

5.3 *Typical independent blocks for potential underground space project*

Independent open land is an important way to meet the shortage of functional supply in the surrounding areas by underground space. This plan aims at the problem of large-scale campus of THU and obvious traffic distance and shortage of some campus functions caused by over--functional zoning. However, due to insufficient green field shortage, the allowable building construction land is limited. We choose the suitable underground space with less influence on the aboveground space as the supplementary part. The necessary function of the functional area is in short supply. Exemplified by three main sports fields of THU campus as typical plots.

– Western Playground. The parking demand of surrounding canteens, gymnasiums, stadiums, academies of Humanities and social sciences, medical schools, libraries and staff dormitories is large and scattered. Surrounding is historical feature protect areas, with insufficient and crowded parking. At the same time, indoor sports venues need to be supplemented. Therefore, it is possible to develop long-term playground to develop the Western playground. It is considered to reserve motor vehicles' entrances and exits, air outlets and personnel import and export in the open space around the playground. Two storehouses and one sport hall can be built underground. The building area can reach 30,000 m², with 400 parking and 10,000 m² sports field.

– Eastern Playground Areas. The surrounding gymnasium and sports ground are large and the teaching staff canteen and human library is near. The demand for parking is scattered but the total demand is large. Many sports fields is suitable for develop underground space. Various combinations can be selected for the underground garages, while building indoor stadiums to improve the quality of sports and leisure space for teachers and students, but also to provide amateur services for social peoples. Optimized C and F plots, each providing three-story underground parking and recreational areas, a total floor area of 25,000 m², each parking 300 vehicles, sports space about 7000 m².

– Bauhinia Playground Areas. The exclusive student apartment area in this area has a large number of playgrounds and large area. In the long term, it should preserve the large area of track and field, preferring G plot (high terrain, large area, good mouth conditions), as the main body of the new student center, as well as underground parking and indoor sports venues to supplement. The parking lot is linked to the long-term urban planning road. It can provide three floors of underground buildings, with an area of about 45,000 m² and a parking lot of 400 vehicles. The underground student center is 20,000 m².

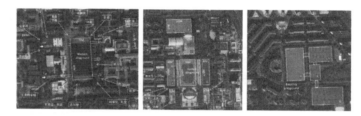

Figure 8. Proposed plan of typical independent underground space potential blocks of THU main campus.

6 CONCLUSION

This paper not only introduces the feasibility and demand of underground space use at THU main campus, but also shows the potential of compound development of underground space for integrated campus-urban space use beneath the campus.

REFERENCES

Zhu, L. & Miao, R.X. 2011. Blueprints of Tsinghua University in the Past One Hundred Years. *Art & Design* 217(05):22–29. (In Chinese)

Luo, S. 1984. Architectural planning evolution of Tsinghua University campus: 1911–1981. *New Architecture* 1984(4):2–14. (In Chinese)

Venturi, Scott Brown and Associates, Inc. 2005. *CONSULTING SERVICE FOR TSINGHUA UNIVERSITY CAMPUS PLANNING REVISIONS.* China: Beijing.

Author Index